PRAISE FOR *BELIEVING IS SEEING*

I can see many exclaiming, "This is the book I've been searching for!" Dr. Michael Guillen has written an exceptionally accessible overview on how science and faith are not only compatible but also infinitely parallel and wholly inseparable. Parents, this is the book to give to your science-minded sons and daughters before they go off to college.

> **JAMES M. TOUR,** Professor of chemistry, materials science and nanoengineering, and computer science, Rice University

Michael Guillen is that rare breed of science journalist who is also an actual scientist. Not surprisingly, then, *Believing is Seeing* is as deep and profound as it is readable.

> **KARL E. JOHNSON,** Founder, Chesterton House, and executive director, Consortium of Christian Study Centers

In *Believing Is Seeing*, Michael Guillen reveals how magnificently and inextricably science and the Christian faith are intertwined. The same pervasive beauty, elegance, logic, and truth permeate both. This awe-inspiring book comes from one of the most colorful and engaging and laugh-out-loud entertaining communicators I have ever met.

> **HUGH ROSS,** President of Reasons To Believe and author of *The Creator and the Cosmos* and *Weathering Climate Change*

This is a fascinating, readable, and compelling book, especially for those who think our wonder-full universe is one grand accident. It isn't.

> **ROBERT C. FAY,** Emeritus professor of chemistry, Cornell University

In this striking account, the author explains why he gave up his atheism to put his faith in God. With great skill, important concepts in contemporary physics are introduced for the nonspecialist, and the author shows how such ideas played a key role in his gradual journey to faith. Science, he explains, used to be his god—but no longer. Yet the author remains a passionate advocate for science, albeit now appreciated within a much broader worldview. The style is fast-paced, and there are plenty of fascinating facts and anecdotes, some quite personal, to keep the reader fully engaged until the last page.

DENIS ALEXANDER, Emeritus founding director of the Faraday Institute for Science and Religion and emeritus fellow of St. Edmund's College, Cambridge, UK

What awaits you in these pages is an exhilarating discovery of a worldview that will enable you to confidently navigate the inevitable storms that lie ahead *and* every challenge, disappointment, and setback you're facing right now. Wherever you are on your journey of faith—deeply steeped or hard-boiled skeptic—*Believing Is Seeing* will jolt you from your comfort zone, propel you toward finding and discovering Truth, and bolster and deepen even your tepid faith. A must-read book for our times.

NANCY STAFFORD, Actress (*First Lady*, *Matlock*), speaker, and author of *The Wonder of His Love* and *Beauty by the Book*

It is unusual to find a world-class scientist who is also gifted at communicating complex ideas. Dr. Michael Guillen is one of those rare gems. In his latest work—destined to be a classic—Dr. Guillen puts these skills to good use as he offers us his insight into the nature of reality. Guillen helps us see that all facets of our lives depend on our faith commitments, on what we believe—even logic, math, and science. This insight means that, despite popular opinion, there isn't a conflict between faith and reason. Instead,

Guillen makes a compelling case that reason rests on a foundation of faith. After reading *Believing Is Seeing*, you will be compelled to examine your own beliefs. Does your faith rest on what is true? Is it a misguided faith—or an enlightened one?

FAZALE "FUZ" RANA, PhD, Biochemist, and author of *The Cell's Design*, *Creating Life in the Lab*, and *Humans 2.0*

I remember a time when my son was embarrassed about asking questions in school, and I told him the smartest people ask the most questions. Dr. Michael Guillen is one of the most brilliant men I've ever met. He is curious about everything and never stops asking questions and helping others find the answers. In his new book, *Believing Is Seeing*, he brings together science and the Bible and shows us how to create a worldview based on absolute truth.

ROBERT MORRIS, Senior pastor, Gateway Church, and author of *The Blessed Life*, *Beyond Blessed*, and *Take the Day Off*

No one discusses the relationship between IQ and SQ (spiritual intelligence) better than Michael Guillen, encouraging a deep reflexive response. This book needs to be read by anyone critically examining the role their individual faith and belief system plays in guiding them toward purpose, service, and leadership.

STEPHEN KIRNON, Program chair social entrepreneurship and change, Pepperdine University and serial life-science entrepreneur

What I particularly appreciate about Michael is his passion for the discoveries of both science and Christian faith and the way he combines these with a truly impressive academic training and a curious and brilliant mind. All this (and more) you'll experience within these pages.

GREG COOTSONA, Author of *Mere Science and Christian Faith*

In *Believing Is Seeing*, physicist and science journalist Michael Guillen tells the story of how he discovered a deep consonance between the scientific world picture and a Christian worldview. He not only shows how the presuppositions that make science possible flow naturally from belief in God, but he also shows that scientific discoveries about the design of life and the universe provide evidence for the reality of God. Guillen's engaging intellectual autobiography will help people from all walks of life wrestle with, and answer, life's deepest questions. Highly readable and highly recommended!

STEPHEN C. MEYER, PhD, Director, Center for Science and Culture Discovery Institute, and author of *Return of the God Hypothesis*

Many think of faith and science as polar opposites. Yet, in *Believing is Seeing*, Michael Guillen shows how, at its root, science itself stands on a foundation of faith. With the insight of a trained astrophysicist and the easy-to-read clarity of a journalist, Guillen writes about physics, astronomy, mathematics, and the methods of science as a whole, showing how they rest on truths that must be accepted by faith.

As a boy growing up in the barrios of Los Angeles, Guillen dreamed of being a scientist. He made it from the barrio to the halls of Ivy League universities and to an Emmy award–winning career in television. But the most surprising turn in his story happened when his study of nature led him to a surprising encounter with the God of nature. In telling his story, Guillen challenges us to examine our own story—and how that story relates to our view of the world around us.

I recommend this book for those who want to know if it is possible to resolve the conflict between science and faith and for those interested in exploring the deep truths of life to which science points.

EMMANUEL HAQQ, PhD, Senior pastor, Christ Community Church, Belchertown, MA

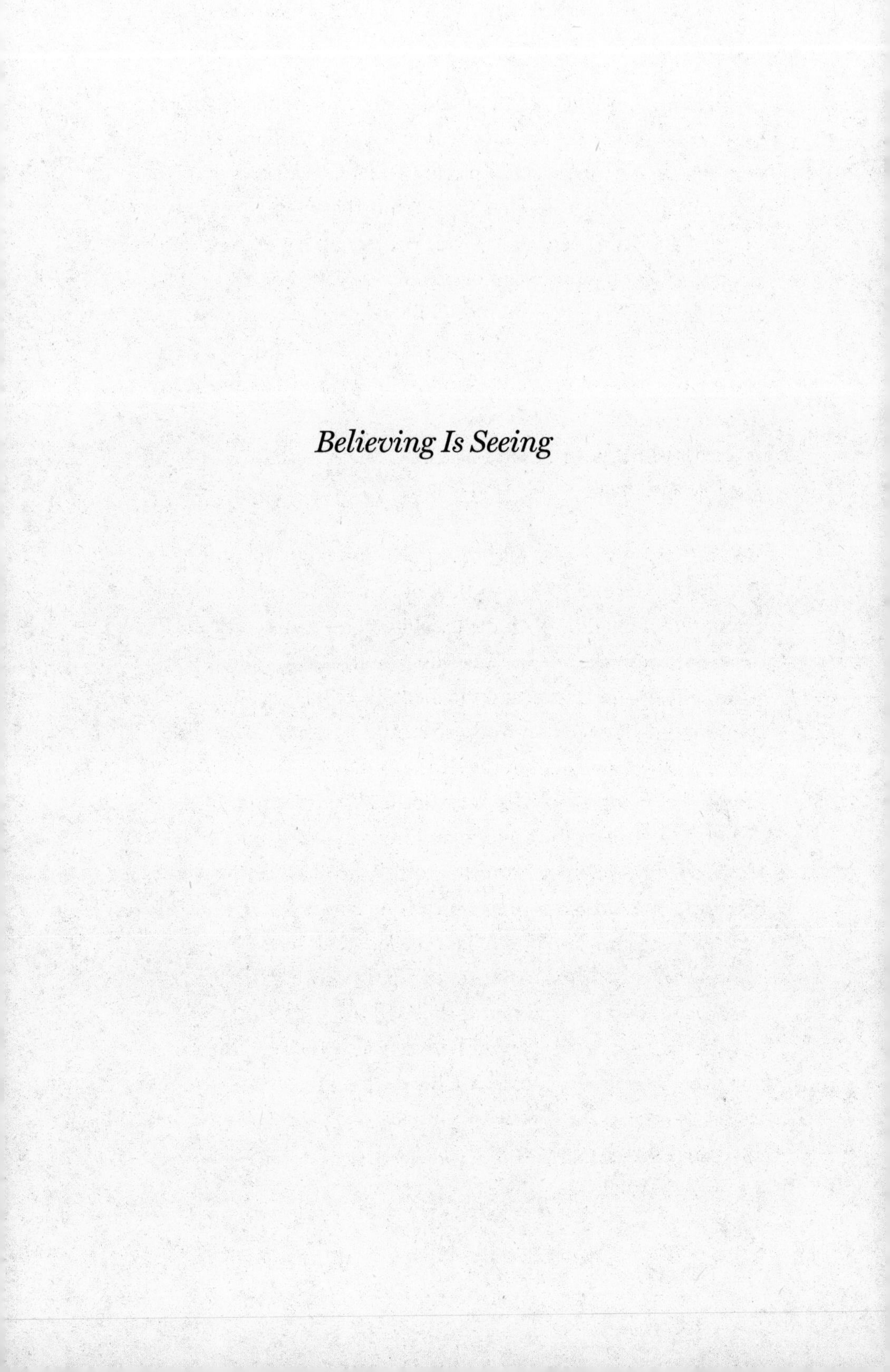

Believing Is Seeing

A Physicist Explains
How Science Shattered His Atheism
and Revealed the Necessity of Faith

BELIEVING IS SEEING

MICHAEL GUILLEN, PhD
FORMER ABC NEWS SCIENCE EDITOR

TYNDALE
REFRESH™
Think Well. Live Well. Be Well.

Visit Tyndale online at tyndale.com.

Visit Tyndale Momentum online at tyndalemomentum.com.

Believing Is Seeing: A Physicist Explains How Science Shattered His Atheism and Revealed the Necessity of Faith

Designed by Faceout Studio, Tim Green

Edited by Dave Lindstedt

The author is represented by Ambassador Literary, Nashville, TN.

For information about special discounts for bulk purchases, please contact Tyndale House Publishers at csresponse@tyndale.com, or call 1-855-277-9400.

Library of Congress Cataloging-in-Publication Data

A catalog record for this book is available from the Library of Congress.

ISBN 978-1-4964-5557-4 (HC)
ISBN 978-1-4964-5558-1 (SC)

Printed in the United States of America

27 26 25 24 23
7 6 5 4 3

ALSO BY THE AUTHOR

Bridges to Infinity: The Human Side of Mathematics

Five Equations That Changed the World: The Power and Poetry of Mathematics

Can a Smart Person Believe in God?

Amazing Truths: How Science and the Bible Agree

The Null Prophecy (thriller novel)

The End of Life as We Know It: Ominous News from the Frontiers of Science

FOR DR. RICE BROOCKS

Precious friend and colleague,

who has dedicated his life to Truth

and understands that it is

far bigger than proof.

CONTENTS

INTRODUCTION

Why I Wrote This Book

As a physicist, mathematician, astronomer, and Christian, I have a worldview broad enough to accommodate both the scientific method and the Bible . . . reason and faith . . . the universe and God.

During my long and winding intellectual and spiritual journey, I've learned two enormously helpful lessons. First, logic does not represent the pinnacle of human intelligence, critical thinking, or wisdom, and it's not faith-free. Second, science is not the enemy of God; instead, it is God's *gift* to humanity, a brilliant way to explore his transfinite nature and stunning creation.

For the past several years, I have been touring university campuses far and wide, answering students' questions about science and Christianity—a hot-button issue since at least the days of Darwin. The questions span the full gamut of human curiosity, from "Do you really believe the *entire* Bible?" to "Do you think that science can explain *everything*?"

Wherever I go—be it Reykjavik or Warsaw or New York City or Phoenix—my young audiences include fresh-faced, impassioned Christians, Atheists, New Agers, Muslims, Buddhists, Nones, you name it. Typically, they keep me up past midnight, thirsting after answers about logic and faith, science and religion, exceptionalism and pluralism—worrying about what the future will look like for them, individually and collectively.

From these wonderful face-to-face meetings, I've made lots of young friends and learned many things about their emerging—and in many ways unparalleled—Gen-Z generation. One lesson is crystal clear: The traditional Christian Church has lost young people by the tens of millions, even those who were brought up by devoutly Christian parents. As many young Christians go off to college and find themselves surrounded by vocal skeptics, they are tempted to believe that God, Jesus, and the Holy Spirit are childish fables and the very idea of faith is somehow lowbrow.

Away from home, feeling set adrift, and beset by uncertainty, many of these young people have turned to science as their go-to authority. And no wonder. They've grown up reading about the miracles wrought by science and technology: connecting the world through handheld devices, creating humanlike robots, curing diseases, sending spacecrafts to far-off worlds, decoding the human genome, inventing new forms of life, even restoring sight to the blind.

But make no mistake: Science, too, has alienated many young people. More than any other generation, Gen-Zers are suffering the unintended consequences of the Age of Social Media. They're watching with dismay as their peers are devastated by unprecedented levels of depression, loneliness, and suicide. In fact, the very night I was speaking at the University of Kentucky, a student in one of the dorms took his own life.

And it's not just young people. I dare say there isn't anyone today who isn't concerned, for example, about the Internet of Things (IOT), that burgeoning network of Web-controlled "smart" gadgets that rule our lives—from voice-activated assistants and TV sets to coffeemakers and vacuum cleaners. Or the double-edged scientific innovations that now threaten our human identity, livelihoods, and privacy—such as genetic engineering, artificial intelligence, and facial-recognition technology. Some distraught rebels have tried "cutting the cord"—shunning social media, disconnecting from the IOT, or living off the grid—but the devastating head-on collision between freedom and technology demands deeper, wider, more thoughtful remedies than that.

It might even require a complete overhaul of your current worldview.

What do I mean by *worldview*? It is how you see yourself, others, the cosmos, and God. It is your religion, whether you call it that or not. It is your own personal Svengali, the all-powerful puppeteer deep within your subconscious that pulls your strings, that controls not only how you see everything but how you react to everything—including this book.

You might pride yourself on being a smart, sophisticated, modern-day person with a smart, sophisticated, modern-day worldview, but don't kid yourself. Your worldview is not based on logic. It is based on faith. That's right: *faith*. Like everyone else's worldview—including mine—yours is ultimately based on what you *believe* to be true, on ideas and feelings that cannot ever be proved.

That is why faith, far from being a weakness, is far stronger than logic, stronger than empirical evidence, and certainly stronger than feelings. Faith is the mysterious, widely misunderstood agent that powers every one of your worldview's unprovable beliefs. It's the granite foundation that supports the entire weight of your worldview.

Faith dictates how you see, think about, and relate to everything within and beyond the universe. Everything. In other words, *believing is seeing.*

Atheists boast about being "free thinkers," but they rely on faith every bit as much as a fundamentalist Christian does. Every one of an Atheist's so-called free thoughts is based on assumptions that cannot be proved. We'll get into that in a later chapter.

Likewise, people calling themselves "true believers" boast about going through life relying purely on faith. Yet, when pressed, they cannot explain the actual phenomenon of faith. In this book, you'll discover what it is, exactly.

On top of all that, many individuals—and you might be one of them—see logic and faith as implacable adversaries and, therefore, believe they must choose between the two. Worst of all, they talk up the importance of *evidence-based thinking* but have a woefully fractured view of what it is.

A recent global survey gave high school students a reading assignment and tested their comprehension. On average, fewer than 9 percent of 15-year-olds surveyed were able to tell the difference between fact and opinion.[1] *Nine percent!* Worse, this appalling derangement doesn't afflict only teenagers; we see it in full display in today's professional media, where editorializing and propagandizing are routinely presented as factual reporting.

The results of this poll and others affirm what I've learned firsthand from my speaking tours: Not only do many of today's young people equate opinion with fact, but they also believe that opinions and feelings are *more important* than facts and that *faith* is like an ugly four-letter word.

This pervasive malady is bad news for both science and religion. The worst of all worlds.

The personal and social fallout from such a severely misguided worldview portends a grim future for you, me, and our loved ones. For the United States. For the planet.

Indeed, if things don't improve, today's unprecedented levels of loneliness, depression, and suicide will prove to be only a preview of greater tragedies to come. When life hits the fan, people young and old will discover too late that their unexamined worldviews are empty, toxic, and even deadly.

So then: *What is your worldview?* Have you ever thought about it? Most people haven't.

If you're like most people, your worldview is like your car. You give it just enough attention to keep it running. You take it into the shop when something's amiss, but you have never once crawled underneath the chassis or looked under the hood to see how it works.

Even now, despite what I've already told you, you might be wondering: *Honestly, does my worldview really matter that much?*

Yes, it does.

It's why I've written this book.

This book teaches you how to become your own worldview mechanic. Using real-life stories, I show you how to hoist your worldview onto a hydraulic lift and inspect it from the bottom up. I show you how to give it a tune-up—or if necessary, a complete overhaul. So when you're done, you'll drive away with a big, beautiful, new, truth-based worldview that will serve you well in life—especially in times of trouble.

In these pages, you'll read about what happens to people who fail to inspect their worldview. People in crisis whose dysfunctional worldview actually ended up killing them. People whose misguided Svengali shone their problems in the worst possible light—causing them to lose hope, to believe there was no conceivable way their lives could get any better.

What about you? What personal setbacks are you facing right now? Is it your health? Your job? A relationship with someone near and dear to you: friend, relative, spouse, lover, child?

Whether you realize it or not, in times of trouble, your worldview is your most treasured possession—arguably your *only* treasured possession. Why? Because it's pulling your strings. It's controlling how you see and react to whatever is vexing you.

Is your worldview big enough to include God? Is it based on enlightened faith? On misguided faith? Depending on the answers, your worldview will either rescue you from your troubles or send you over the edge. It will ultimately spell joy or continued sorrow. Life or death.

That's what's at stake for you. *That's* why your worldview matters so much. And *that's* why your very next decision will determine which fork in the road you'll take on the journey of life.

Will you stop reading here, put down the book, and drive away in your old, beat-up worldview? Or will you forge ahead and learn how to claim a brand-new, well-informed, powerful worldview? One that will see you safely and serenely through the inevitable tempests of life—and more than that, fly you to the highest possible, most breathtaking mountaintops of human experience.

The choice is now upon you.

The very next instant is the beginning of your future.

MY JOURNEY

What mind can penetrate your nature?
What language can express this marvel? None, to be sure.
This is where human discourse turns toward
the contemplation of the divine.

LEONARDO DA VINCI

1

CALIFORNIA DREAMIN'... AND BEYOND

Ah, but a man's reach should exceed his grasp,
Or what's a heaven for?

ROBERT BROWNING, "ANDREA DEL SARTO"

Three things you should know about me: First, I was born in the heart of East Los Angeles—I'm five-eighths Mexican, one-quarter Cuban/Spanish, and one-eighth Austrian (go figure)—and I *love* Mexican food. Second, I'm a bit of a rebel. Ever since I was young, I've gone my own way, even to the point of being deliberately contrary at times. Third, I'm curious about *everything*. I drove my teachers crazy with all my questions. In junior high, my math teacher hated me because I constantly interrupted his class. Finally, he got so fed up that he started calling me Michael *Jillion*—because I asked a jillion questions. But I didn't care. I kept asking.

In some ways, I relate strongly to Dr. Frankenstein—though not in the sense of piecing together dead body parts and making them come to life! What I resonate with most is his passion for wanting to know how the universe works.

In Carl Laemmle Jr.'s 1931 film adaptation of the Frankenstein story—my absolute favorite version—the role of the rebel scientist is played by Colin Clive, a suave-looking British actor. Just after creating the monster, he is scolded by the elderly, straitlaced Dr. Waldman, played by Edward Van Sloan. In my opinion, this iconic scene captures perfectly the heart and soul of a scientist:

> **DR. WALDMAN:** "This creature of yours should be kept under guard! Mark my words: He will prove dangerous."
>
> **DR. FRANKENSTEIN:** "Dangerous? Poor old Waldman. Have you never wanted to do anything that was dangerous? Where should we be if nobody tried to find out what lies beyond? You never wanted to look beyond the clouds and the stars, or to know what causes the trees to bud? And what changes the darkness into light? But if you talk like that, people call you crazy. Well, if I could discover just one of these things—what eternity is, for example—I wouldn't care if they did think I was crazy."[1]

In second grade, I began to dream—*literally*—of becoming a scientist. In my sleep at night, I saw myself wearing a white coat, working in a lab filled with equipment, and being awarded the Nobel Prize—for what, I can't remember. All I know is I was happy as a pig in slop.

Chasing that blissful dream got me out of the barrio and into UCLA, where I earned a BS in physics and mathematics. Afterward, I applied to and was accepted by the graduate physics department in several famous universities. Cornell was one of them.

Before I made my decision, my dad and I started out by flying to Upstate New York to visit the Cornell campus. It was late March, and the trees had no leaves. We both thought there had been a fire. Growing up in Southern California, we'd never seen anything like it.

I met the physics faculty and toured the Wilson Synchrotron Laboratory, a world-class atom smasher right there on campus. I was scheduled to visit Princeton next, but I told my dad it wasn't necessary. I was convinced Cornell was the perfect school for me.

The morning of our departure for home, we were awakened by a phone call from David Cassel, the physicist who was to become my thesis chairman.

"Morning!" he chirped. "Have you looked outside yet?"

"No," I replied, rushing to open the blinds on the big window of our hotel room, revealing a landscape covered in snow.

"Welcome to Ithaca!" Professor Cassel sang out.

After checking out of the hotel, Dad and I couldn't resist. We stepped outside and immediately started playing in the snow like a couple of overgrown kids. The lady behind the front desk stared at us incredulously, and for good reason. As I was to learn, by late March, Ithaca residents were well fed up with snow.

Several months later, when I returned to Cornell by myself to begin my studies, I felt as if I'd won the lottery. This poor little nobody from the wrong side of the tracks was actually going to become a physicist. *Imagine!*

It was the beginning of a brand-new and *very different* life than I'd had up till then.

I was reared in a strict, Spanish-speaking, Pentecostal household. My dad and both my grandfathers were ministers. In fact, for four decades, my paternal grandfather, after whom I am named, was the greatly adored president of *Concilio Latino Americano de Iglesias Cristianas* (CLADIC), the oldest, independent, Spanish-speaking Pentecostal organization in the nation, comprising churches in the US, Mexico, and Central America.[2]

When I was growing up, my family attended church every day; and the services were long, drawn out, raucous affairs. I remember the entire congregation, including my mother, jumping up and down and ecstatically speaking in tongues. CLADIC members were forbidden from dancing, watching television, and a host of other things considered mentally, physically, and spiritually unhealthy.

The Bible claims that a blessing is a divine inheritance passed from generation to generation. So everyone I knew expected me to become a minister—and perhaps even one day succeed my grandfather as president of CLADIC.

But I was completely devoted to science, not to church services and what I considered ancient, supernatural beliefs. Even though I lived in a strict Pentecostal home, my mind, attention, and curiosity were somewhere else entirely. I was captivated by numbers and logic, natural phenomena, and the scientific method. And little by little, I absorbed the scientific worldview until it became my own. By the time I graduated from UCLA, I belonged body, mind, and soul to the worlds of science and Atheism, which seemed to me to go hand in hand.

When I departed LA for Cornell, therefore, I was very sad to say good-bye to my family and friends, but I was more than happy to leave behind the religion I had never really embraced. I was also relieved to escape from the pressure of going into the ministry, something I had *zero* interest in doing.

In short, the experience was *liberating*!

When I arrived in Ithaca and it began to sink in that I didn't know a solitary soul there, I realized it was fine with me. More than fine, actually, because it underscored that I was starting a brand-new life. *My* life. *My* dream. The dream I'd worked so hard to attain. The dream of becoming a monk. A *scientific monk*.

Fueled by passion and more than a little caffeine, I spent my days and nights either in class or in a dungeon-like lab—just like Dr. Frankenstein! At most, I slept maybe three hours a night, typically from three to six o'clock in the morning.

My lab was in the basement of Cornell's high-energy physics

building, the Laboratory of Nuclear Studies (LNS).[3] Inside that spacious, windowless man-cave, I couldn't tell whether it was day or night, and I didn't care. I barely ate, and when I did, it was mostly from vending machines in and around the LNS. Truth be told, I was a skinny, unkempt, intense uber-geek, sporting tight corduroy jeans and a cloud of unshorn curly brown hair.

I had no social life, no friends to speak of, and my family was 2,700 miles away. But I was perfectly content. All I cared about and thought about was *science*.

At first, as a twenty-one-year-old freshman grad student, my curiosity was laser focused on learning what the universe was made of. What were its most fundamental elements?

I got to thinking: When you magnify a digital photo, you see pixels, right? So, then, if you magnify the universe—push past its electrons, protons, neutrons, quarks, gluons, and so forth—if you keep magnifying and magnifying, what will you ultimately see? Pixels of matter? Pixels of energy? Pixels of spacetime? I was more than eager to find out.

One day, however, a group of observational astronomers, led by Princeton's legendary P. J. E. Peebles, announced that galaxies are not scattered randomly throughout the universe, as we'd always supposed. Rather, they form a pattern, like a magnificent 3D work of art.

Where did this pattern come from? What did it signify? Was it just an accident?

Suddenly, *those* were the deep questions I wanted to answer. But it would mean switching from focusing on *pixels*, the smallest things in the universe, to focusing on *galaxies*, the largest things in the universe.

Changing your major in grad school isn't easy, but I didn't care. I was determined to follow my own path. Told that I needed permission from Hans Bethe, Cornell's legendary theoretical physicist, I went to see him.

In the 1940s, Bethe had led the theoretical division of the Manhattan Project, which created the world's first atomic bomb. In the 1960s, he won a Nobel Prize for explaining why the sun shines.

Bethe was an old-school, no-nonsense German hard-liner, whose office was on the top floor of the LNS. We grad students were afraid of him—*and* of Velma Ray, his formidable secretary, whom we had to get past in order to see him.

It didn't take long for Bethe to decide my fate. In his thick German accent, he told me I needed to take two semesters of general relativity, arguably the toughest subject in modern physics. If I did well, he'd let me switch. If not . . . well, I'd have to stick to pixels.

The general relativity courses were taught by Saul Teukolsky, a brilliant young physicist whom Cornell had recently hired away from Caltech. The classwork was challenging, but I passed, and with Bethe's blessing, I began studying galaxies.

ONCE UPON A TIME, IN A GALAXY FAR, FAR AWAY

Very quickly I learned that galaxies rotate slowly, like enormous merry-go-rounds. According to a scientific law called the virial theorem, the more massive the galaxy, the faster it spins.

I also learned that galaxies spin much faster than they should, in apparent violation of the virial theorem. It's as if they are far more massive than they appear—like they're bloated with some kind of unseen material that makes them spin abnormally fast. My astronomy professors called this mystery the *missing-mass problem.*

Today, we call this hypothetical missing mass *dark matter*. Based on what little we know, we speculate it could be an entirely new invisible form of matter, ruled by an entirely new kind of force.[4] But honestly, we don't know *what* it is—or even if it really exists.

More recently, we've discovered another oddity about the heavens that is also totally invisible: *dark energy*. From what we can tell (which is precious little), it behaves like a *repulsive* force that causes the universe to balloon out at an accelerating speed.

And get this: Together, dark matter and dark energy seem to constitute *95 percent* of the entire universe. That's right, scientists now believe that 95 percent of the universe is invisible to us.

When I first learned about the missing-mass problem and what we now call dark matter, it blew my mind, rocked my reality, and challenged my perception of *everything*. (So did the discovery of dark energy, but that happened after I graduated, when I was teaching at Harvard.)

As a pious scientific monk—a liberated, free-thinking Atheist—I lived by the trusty adage that *seeing is believing*. I refused to believe in anything I couldn't actually see and that couldn't be proved.

But that worldview was now out the window because science had discovered that what we are able to "see"—what we're able to prove the existence of—is only a small fraction of what's out there.

The missing-mass problem made me realize that if I stuck with my hard-nosed, scientific worldview—if I insisted that "seeing is believing"—then I'd be turning a blind eye to 95 percent of what's out there in the universe. Clearly, my worldview was too narrow-minded for the cosmos.

It needed some expanding. It had to become big enough to include belief not only in what I could see and prove but in what I could *not* see or prove—such as dark matter. Otherwise, I couldn't honestly continue calling myself a scientist.

BEYOND MY WILDEST DREAMS

As I dove into my investigation of galaxies, I quickly realized I needed to immerse myself in not one, not two, but three different disciplines: physics, astronomy, and mathematics. Once again, I petitioned for permission to make the change.

It was an unprecedented request, but I was fortunate to receive permission—thanks in large part to the unwavering support of my thesis chairman, David Cassel. So I ended up having offices in all three departments, surrounded by three distinct groups of wonderful, brilliant colleagues from whom I learned a great deal.

I remember being very excited when I first learned about kinetic theory. It had always been used to describe the behavior of gases, but I struck on the idea of using it to explain the behavior of galaxies.

I immediately pursued my hunch with the encouragement and mentorship of Richard Liboff, a world-renowned expert in kinetic theory, who ultimately became my thesis advisor.[5] Years later—after one of the most intense, nonstop efforts of my young life—I hit pay dirt. I discovered an elegant mathematical explanation for why galaxies form a spectacular 3D pattern in deep space and published my finding in the *Monthly Notices of the Royal Astronomical Society*.[6] The implications of this discovery were potentially revolutionary, so I submitted it for a PhD in physics, mathematics, and astronomy.

I'll never forget the day of my defense-of-thesis exam, the final hurdle I needed to clear before I could receive my 3D doctorate. Inside a small classroom on the top floor of the LNS, I stood at the blackboard facing professors from all three disciplines. According to the rules, they were allowed to ask me any questions they wanted, no matter how tough. And, sure enough, they let me have it!

The exam lasted four grueling hours, but I passed! And I'm not ashamed to admit that I wept like a baby as, one by one, my committee members offered a handshake and said, "Congratulations."

My dream had finally come true! I could not imagine being any happier!

Little did I know that, just a short while later, on my way north to Harvard, an even greater, more exciting adventure awaited me—one I never could've dreamed of. For as I'm now fond of saying, "A funny thing happened to me on the way to Cambridge."

On the trip, I stopped at the Museum of Natural History in Washington, DC, to attend a seminar on George Orwell's novel *1984*. It was hosted by Fred Graham, then the legal correspondent for CBS News.

At the reception afterward, I saw Graham and a woman standing alone, so I introduced myself to them. When Graham found out I was a scientist, he said something like "Hey, maybe you can settle an argument I'm having with my producer here."

"Sure," I said. "What's the problem?"

"You know that giant pendulum out in the rotunda? My producer says that once you get it going, it'll never stop swinging. I don't agree," he said. "I think you need to push it now and then to keep it going."

For me, it was a no-brainer.

"It's called a Foucault pendulum," I explained. "And there isn't much friction to slow it down—just a little rubbing where the steel cable attaches to the ceiling. But it's enough to gradually slow it down; so, yes, it *does* need to be nudged now and then."

Graham leapt at my explanation. "Wow!" he said. "Would you like to be on television?"

I thought he was kidding.

"No, really," he said. "CBS News is looking for a science reporter. If it's okay, I'd like to nominate you. I *love* how you explain things."

I went on to Cambridge, hardly believing what had just happened. I started my teaching job and soon began doubting that anything would come of the Graham encounter. But sure enough, weeks later, *CBS Morning News* hired me to be its new science and technology correspondent.

I was assigned to work with a veteran New York–based producer named Gail Eisen; years later, she went on to produce Diane Sawyer at *60 Minutes*. Gail patiently and expertly taught me the ropes, and in no time, I found myself appearing regularly on national television.

At Harvard, meanwhile, I had the honor of teaching under the senior leadership of Roy Glauber, a physicist who later won a Nobel Prize for a discovery he made in quantum physics. I loved teaching (and still do), so I was greatly moved when I was twice awarded Harvard's prestigious Danforth Award for Excellence in Teaching.

After appearing on *CBS Morning News* for a few years, I was stolen away—first by Phil Balboni, the famous news director at WCVB, the ABC affiliate in Boston, and then by ABC News itself, based in New York City.[7]

At first, I did science reports only for *Good Morning America*. But soon enough I also began appearing on *Nightline*, *20/20*, and *World News Tonight*. Altogether, I had the great honor of working with Barbara Walters, Hugh Downs, Ted Koppel, Peter Jennings, Joan Lunden, Diane Sawyer, Oprah Winfrey, Connie Chung, and many other first-rate professionals.

During those years, I split my time between Harvard and ABC News. It was a fun, glamorous life—but also a tumultuous and stressful one. One day I'd be on campus, inside Harvard's Science Center, teaching physics to undergraduates. The next day I'd be flying to Japan to cover a volcanic eruption. Or to Alaska to cover an oil spill. Or to the South Pole to cover the ozone hole. Or to the North Pole to cover the first transarctic dogsled expedition. Or to England to interview Stephen Hawking. Along the way, I won three Emmys and became the first person to broadcast live to North America from the Antarctic and the first television correspondent to travel to the bottom of the Atlantic Ocean and report from the wreck of the *Titanic*.

In 1994, after nine magical years, I reluctantly left Harvard to work full-time in television. It felt weird to no longer be affiliated with an academic institution, but I was tired of leading a hectic double life between the classroom and the studio.

Eventually, after fourteen thoroughly enjoyable years, I also left ABC News. My wife and I were planning to have a family, and we both agreed that my being a jet-setting news correspondent was incompatible with being a good father.

Shortly afterward, the History Channel hired me to host a weekly, prime-time series called *Where Did It Come From?* And later, the John Templeton Foundation gave me a large grant to produce a full-length motion picture celebrating human generosity. That movie, *Little Red Wagon*, won many awards.

To put it mildly, my life did not turn out the way that dreamy little Mexican kid from East LA could ever have imagined. What's more, as you're about to see, the unexpected twists and turns I've just described were just the tip of the iceberg.

2

THE AWAKENING

There are more things in heaven and earth, Horatio,
Than are dreamt of in your philosophy.

WILLIAM SHAKESPEARE, *HAMLET*

During my scientific training at Cornell, I learned that black holes are like mysterious Bermuda Triangles of outer space; anything near them is sucked in and disappears forever. I learned that virtual particles are subatomic ghosts that inhabit a quantum-mechanical Twilight Zone: They're real but not real. I learned that space and time are elastic; depending on the reference frame, they expand and contract, like Elastigirl and Mister Fantastic.

I learned that galaxies in deep space are not randomly distributed but are arrayed in an elegant three-dimensional pattern. I learned that the universe is expanding like crazy, implying that everything in it—mass, energy, space, time—is the aftermath of a cosmic-sized explosion. I learned that 95 percent of the "observable" universe is invisible to us and that beyond it is an *unobservable* universe that is *100 percent* invisible to us.

All of this took my breath away!

Pretty soon I began asking myself a simple but profound question: *How did this amazing, mostly invisible universe of ours come to be so amazing and mostly invisible?*

My areas of expertise—physics, astronomy, and mathematics—offered me an answer: The universe was created accidentally by a random perturbation (or disturbance) in the quantum vacuum (QV). Though it's said to be completely empty, the QV is actually filled with unseen force fields, which at any time and without warning can cough up real, honest-to-goodness atomic particles. Think of the invisible quantum fields as phantom guitar strings that produce audible notes when plucked hard enough.

In other words, science tells us that, at some random time in the past, someone or something plucked the invisible force fields within the supposedly empty QV and out came the sound of music—the makings of a vast physical universe.

This admittedly intriguing hypothesis required me as a grad student to believe in a flabbergasting paradox: The QV is absolutely *nothing* (which is why we call it a vacuum), but it also has the potential to be absolutely *everything*—like a womb that is simultaneously barren *and* fertile.

If that paradoxical notion sounds supernatural, it is. But it's backed by quantum physics, a highly valued crown jewel of modern physics. So, despite its seeming mysticism, we scientists take it very seriously.

We even have a fancy, smart-sounding way of describing the contradictory nature of the QV. We say that *nothing is unstable.* According to this hypothesis, if we wait long enough, *nothing* will inevitably become *something*—even an entire universe.

For good measure, we have given this fantastic idea a sober-sounding scientific name: *the standard model of cosmology* (SMC). It's also known as the standard model of big-bang cosmology or (for technical reasons I won't get into here) the lambda cold dark matter model, just to name a few.

For a long time, like a good scientist, I went along with the crowd. *Yes,* I told myself, *the SMC completely answers the question: How did this amazing, mostly invisible universe of ours come to be so amazing and mostly invisible?*

Even today I eagerly point out that the SMC is backed by compelling physical evidence and by equations that I find incredibly beautiful—such as these:

$$R_{\mu v} - \frac{1}{2} R\, g_{\mu v} + \Lambda\, g_{\mu v} = \frac{8\pi G}{c^4} T_{\mu v}$$

$$H = \sum_j E_j (c_j^\dagger c_j + 1/2)$$

If you're not a math person, don't be put off by the equations in this book; simply admire them as you would works of art—as dazzling evidence of God's brilliant design.

But I eventually learned that the SMC has some serious problems—problems that have only gotten worse, not better, since I was a grad student. In fact, cosmology today is in a state of crisis.

Let me explain.

Broadly speaking, the SMC is the marriage of two towering intellectual achievements: general relativity (GR) and quantum physics (QP).

General relativity is our best scientific explanation for the gravitational force and all the cosmic-sized objects it manipulates: planets, stars, solar systems, galaxies, galaxy clusters and super-clusters, quasars, pulsars, fast radio bursts, supernovae, black holes, white holes, and so forth.

Quantum physics is science's best explanation of the three other known forces—electromagnetism, weak, and strong—and all the atomic- and nuclear-sized objects they push and pull: chemical elements, electrons, protons, neutrons, quarks, gluons, and on and on.

The first beautiful equation I referenced above is Einstein's field equation, the centerpiece of general relativity. The second equation describes the quantum vacuum, a central concept in quantum physics.

So what's the crisis?

The two equations—and the two theories themselves, GR and QP—are fundamentally *incompatible*. As inimical as fire and water. Which means their marriage within the SMC is a disaster. A deception, really.

Let me put it a different way.

In 1994, I covered the Chunnel story for ABC News. The Chunnel is a 31.5-mile tunnel (actually three parallel tunnels) running beneath the English Channel with high-speed trains connecting roughly Folkestone, England, and Coquelles, France.

Using giant boring machines, English and French engineers dug the tunnel from opposite ends. It took meticulous planning, ingenuity, and a clever use of GPS technology to make sure the two efforts married up perfectly in the middle.

The crisis in cosmology stems from the fact that the two halves of the SMC—GR and QP—do *not* line up. GR and QP are like a pair of enormous boring machines chewing away from opposite ends of the universe: the cosmic and the nuclear. Each machine is terrific at what it does, but there is a catastrophic disconnect between their efforts.

For decades now, we've sought—and continue seeking—clever ways of unifying and harmonizing the dysfunctional marriage between GR and QP. Proposed therapies include various *string theories*, which claim that the pixels of spacetime are not four-dimensional points, as GR sees them, but strings of many dimensions.

That singular idea, which I find quite intriguing, appears to fix the disconnect between GR and QP. It also leads to the astonishing possibility that our universe is but one of an infinite number of universes. More about this multiverse hypothesis later.[1]

Unfortunately, string theories have their own huge problems. So although they have the potential of solving the existing crisis in cosmology, in some ways they only make matters worse.

As a grad student, I pondered these troubling issues in all their mathematical and technical splendor. During that time, I tried my best to remain steadfastly loyal to the SMC; but ultimately, I was put off by its serious problems.

That left me with a choice to make: Was there another theory that did a better job of explaining the origin of the universe? There were (and still are) many contenders; but each one, I discovered, had its own set of unresolved issues.

It was around this time that Cornell astronomer Carl Sagan was becoming quite famous. He was appearing regularly on the *Tonight Show*, hosted by Johnny Carson, an avid amateur astronomer. Sagan was also creating a worldwide buzz with his search for extraterrestrial intelligence (SETI) and plans to host *Cosmos*, a major miniseries on PBS.

I'd gotten to know Carl and had even taken classes from him in exobiology—classes co-taught by Frank Drake, the equally famous father of SETI. And one thing I picked up on was his fascination with something called the Vedas. He was always talking about them.

One day, probably because I admired Carl so much and was caught up by his growing celebrity status—and frankly also because I was just plain curious—I decided to find out what these Vedas were all about. Since Google didn't exist yet, I went to Olin Library for answers.

I learned that the Vedas are the sacred literature of Hinduism, the oldest religion in the world. Fascinated by this, I dove into Hinduism the way I do anything I'm curious about: with everything I had.

In the process, I discovered and fell in love with the novels of

German author and Nobel laureate Hermann Hesse: *Siddhartha*, *Narcissus and Goldmund*, *Demian*, and many others.[2] I resonated strongly with Hesse's protagonists—tormented intellectuals setting off on long, winding journeys seeking answers to life's deepest questions.

That was me!

In the years that followed, my voracious curiosity got hold of me, and I became fascinated with other metaphysical belief systems as well: Buddhism, Chinese mysticism (including the *I Ching* and Taoism), Islam, and many others. Richard Liboff, my thesis professor, was Jewish, so I also dove into Judaism. I was especially fascinated with the Kabbalah and Jewish mysticism.

As an aside, it was during this worldview-altering journey that I learned to love chopped chicken liver. On Friday nights, Richard and I would attend Shabbat services and then go to his home for a late-night nosh, our own little *Oneg Shabbat*.

The first night, he offered me some chopped chicken liver, which I politely declined. When I was a kid, my mother had constantly fed me fried liver to fatten me up. It didn't work, and I ended up hating the stuff.

Week after week, Richard persisted.

"This is my mother's famous recipe," he always said. "Try it; I guarantee you'll like it."

One Friday evening, I finally gave in. I tried a little bit of the liver on a cracker and—miracle of miracles—instantly fell in love with it.

At one point during my far-ranging intellectual and spiritual quest, I also explored Transcendental Meditation (TM). A famous guru named Maharishi Mahesh Yogi visited Cornell and claimed that sincere TM devotees could levitate.

Imagine defying the law of gravity! As a scientist, I was blown away by the claim.

Unfortunately, it never worked for me. But the experience was fascinating.

BEAUTY AND THE BEAST

One night in the midst of this rambling sojourn—it was probably three o'clock in the morning—I headed to my dorm room for a few hours of shut-eye. When I opened the door, I heard a scraping sound on the floor.

Looking down, I saw a white envelope with my name on it wedged under the door. It was a Valentine's Day card, signed *Laurel.* My first thought was: *I had no idea it was Valentine's Day.*

Laurel was an undergraduate who had taken my Physics for Poets class more than a year earlier. She'd made an impression on me because she sat in the front row and always asked smart questions. Moreover, she was tall, beautiful, and had big brown eyes.

More recently, Laurel had been doing some volunteer work for the Leonardo da Vinci Society, a group I'd founded to encourage multidisciplinary studies—of the very sort I was now undertaking.

But getting a Valentine's Day card from her confused me. She was a popular sorority girl—a member of Kappa Kappa Gamma—and I was an unkempt, monkish geek. It was like stepping into the storyline of *Beauty and the Beast.*

I sought her out and thanked her, and from then on made a radical change in my routine. Laurel lived with some other coeds in a two-story house in College Town with a deck on the roof. Stealing away from my studies late at night, I began going up on the roof with her, sitting under the stars, and talking into the wee hours of the morning.

When I asked Laurel why she'd sent me a Valentine's Day card, she gave me several reasons. I'll mention just two of them.

First, she said, she was intrigued by me. She said that my unorthodox behavior—switching from experimental to theoretical physics, working on a revolutionary thesis that cut across several ultra-rigorous disciplines, founding the Leonardo da Vinci Society, getting by on very little sleep, holing up like a hermit, and so on—had turned me into something of a campus legend. I was like Cornell's very own Phantom of the Opera, she said—except better looking.

Second, despite my reputation as the ultimate science geek, she sensed in me "a latent spirituality." Those were her exact words. She found my exploration of world religions from a scientist's perspective uniquely fascinating because she herself was searching for deeper meaning and purpose in life but, so far, had come up empty.

She told me she had grown up a Catholic, but everything had changed when her parents got divorced when she was fifteen. Her dad had moved out, and her mom—unhappy with Vatican II and enthralled with the feminist movement—had left the Church and plunged into New Age mysticism. Laurel had followed suit.

Laurel had begun attending New Age retreats, where she learned about yoga, aura readings, chakras, crystal power, past lives, and necromancy. At a retreat in Gettysburg, Pennsylvania, she witnessed people claiming to speak to the spirits of dead Civil War soldiers.

At Cornell, she said, she'd dated a nice Christian guy, and the experience had caused her to think again about Christianity. She had sought out the campus priest, but in answer to her probing questions, he'd simply handed her a copy of the *Catechism*.

She asked me what *I* thought about Christianity. I told her about my Pentecostal background and how foreign and prosaic it seemed to me now.

"Have you ever actually read the Bible?" she said.

"No."

Because of my Pentecostal upbringing, I was already familiar with the Bible's basic teachings. So, for me, it was old news. With my limited spare time, I was more interested in exploring exotic religions I knew nothing about.

On top of that, I told her, it was my impression that people who believed in the Bible hated science. That was a big turnoff for me because I loved science more than anything in the world.

Laurel didn't argue with me, but one day, she said: "Hey, I've never read the Bible, either. If you read it, I'll read it with you."

It was an offer I couldn't refuse. Not because I was particularly keen on revisiting a book I felt didn't have anything for me, but because I wanted to spend more time with Laurel.

Little did I know the experience would change my life forever.

WAR AND PEACE

For the next two years—yes, *two years*—Laurel and I read the Bible from cover to cover.

We did most of our reading on Sundays, carving out time from our studies. Usually, we met at Laurel's house—my dorm room was way too small; but sometimes we'd squirrel ourselves away at a secluded table in Willard Straight Hall (the student union) or a local hangout such as Collegetown Bagels.

It took us two years to finish because we picked apart every single word of every single sentence of every single verse of every single chapter of every single book—from Genesis to Revelation. If something didn't make sense, we'd talk it out. We had a jillion questions (of course), and we wrote them all down in a spiral notebook. It was quite a project.

To be honest, we found the Old Testament (OT) to be mostly depressing. God created humanity . . . we messed up . . . he got angry. He kept giving us second chances, we kept messing up, and he kept getting mad. There was no happy ending.[3] In fact, the book of Malachi ended with a threat: "Or else I will come and strike the land with total destruction."[4]

We also found the OT to be extremely *logical*. In that regard, it was exactly like all the other religions I had studied. That commonality really struck me.

According to these very logical religions, we always get what we deserve; karma rules the cosmos and wins the day. You mess up, the universe punishes you. You hurt me, I hurt you. It's always an eye for an eye.

Also, if you want to get into heaven or nirvana or paradise, you gotta earn it. There's no free lunch. The logic is very worldly, very simpleminded, very predictable.

The New Testament (NT), we quickly discovered, was totally different—which was shocking. The whole time Laurel and I were reading the OT, it felt as if we were sitting in the dark. But when we turned the last page of Malachi and started reading Matthew—*wow!*—the lights suddenly came on for me.

In the NT, God isn't mad at us anymore. He wants to make peace with his creation. He does so in a way that's hard to believe but that takes our species by storm and yanks civilization in a wholly new and radical direction.

Laurel and I noticed another huge difference: The NT didn't seem *logical.* For instance, it claims that Jesus is fully a man *and* fully God, which means he's simultaneously mortal *and* immortal. How is that possible?

The things this God-man says also defy logic—such as: You can be a twenty-four-karat gold sinner and still get into heaven; you should love not just your friends but your enemies as well; those who would be first shall be last, and the last shall be first; you must die in order to truly live; the meek, not the strong, shall rule the world.

Huh?

That last claim alone set my mind awhirl because it not only defied logic but is also a main tenet of evolutionary biology. According to Darwinism, the *strongest* and *fittest,* not the meekest, shall always rule the world.

Surprisingly, the New Testament's glaring disregard for logic didn't turn me off. Instead, it reminded me immediately of what I'd been learning in quantum physics (QP).

QP, you see, is decidedly *not logical.* It says things such as:

Something can exist and *not* exist at the same time; something can get from here to there without traveling from here to there; something can be *nothing* and *everything* at the same time.

It all sounds far-fetched—just as the NT does—but there's considerable evidence that QP is trustworthy. It's why we scientists have faith in it.

In other words, QP is not logical, but it's also not nonsense. It mustn't be casually dismissed.

For that reason, I knew I mustn't casually dismiss the NT, either. I needed to take it seriously. I needed to see where it led me. Only then could I reach an intelligent, fair-minded verdict.

This is extremely important. I want to make sure you understand what I'm saying here before we move on.

Just because something isn't logical doesn't mean it's *illogical.* Just because it doesn't make sense doesn't mean it's *nonsense.* If you insist on limiting your analysis to matters of sense and logic, you risk overlooking the most profound truths about yourself and the universe.

Why is that?

Reasoning that doesn't sound logical—that may even, at first blush, sound like utter nonsense—might very well be what I call *trans*logical thinking. Quantum Physics and the New Testament are two powerful examples of translogical thinking.

I first learned about the enormous difference between conventional, logical thinking and translogical thinking while studying QP. Here it is in a nutshell:

- Conventional thinking leads you to **trivial truths**. By "trivial," I don't mean insignificant or unimportant. Rather, a trivial truth is *logical* and *commonplace*. A trivial truth *makes sense.* Above all—and this is key—the opposite of a trivial truth is always *false*.

- Translogical thinking leads to **profound truths**. A profound truth is not logical. A profound truth doesn't make sense. Above all—and again this is key, as well as hard to believe—the opposite of a profound truth is *also* true.

Here's how the renowned Danish physicist Niels Bohr, a cofounder of QP, explained it: "Profound truths [are] recognized by the fact that the opposite is also a profound truth, in contrast to trivialities where opposites are obviously absurd."[5]

Nobel laureate Max Delbrück later paraphrased Bohr: "It is the hallmark of any deep truth that its negation is also a deep truth."[6]

Trivial truths follow the straightforward rules of Aristotelian logic. Profound truths, on the other hand, flout the rules, opening our eyes to the deepest, most amazing, most impenetrable secrets of the universe.

Here are some examples of what I mean.

First, consider this statement:

A DIME IS WORTH TEN CENTS.

This is a *trivial* truth. A conventional truth. Why? Because its opposite is false:

A DIME IS *NOT* WORTH TEN CENTS.

Now consider this statement:

THE QUANTUM VACUUM IS NOTHINGNESS.

This is a *profound* truth. A translogical truth. Why? Because its opposite, by definition, is also true:

THE QUANTUM VACUUM IS *NOT* NOTHINGNESS.

According to QP, the quantum vacuum is both nothing and everything. It consists of empty space but, paradoxically, is seething with countless *virtual* subatomic particles. Like ghosts, virtual particles are said to exist and not exist at the same time.

I know! I know! It doesn't make sense. Like nearly everyone, you're used to thinking logically; you're used to thinking conventionally. You're used to thinking that the opposite of *true* is *false*. So translogical truths—profound truths—don't make any sense to you. But now you know: They are, nonetheless, true. In fact, they are the most profound truths in the universe.

As you'll see as we go on, translogical thinking is the most penetrating way of seeing, probing, and describing the universe. In a sense, it's a superpower unique to the human species; no other animal on the planet is capable of it. Far more sophisticated than mere IQ, translogical thinking is a special kind of intelligence—one that transcends the pedestrian rules of Aristotelian logic.[7]

You are capable of translogical thinking, and so am I. But, tragically, not everyone uses it in daily life—or even knows they have the ability for it.

Back when I was a grad student reading the Bible with Laurel, I'd already had plenty of experience with translogical thinking and profound truths. So I instantly recognized the possibility that the NT was translogical—that, like QP, far from being trivial (in the technical sense of the word that I described above), the NT was signaling profundity. For this reason and others, the NT captivated me as no other sacred literature ever had.

Still, it wasn't enough to put a dent in my Atheism. Not by a long shot.

I'm not that easily swayed.

3

THE RECKONING: PART 1

It's not important to get children to read. . . .
It's much more important to teach children to question what they read.
Children should be taught to question everything.

GEORGE CARLIN

Reading the Bible with Laurel didn't cause me to fall to my knees. I did not experience a Damascus-like conversion.

Instead, I kept asking questions—tough questions—of both my beloved science *and* Christianity.

Most of my efforts centered on one quandary: Is it possible to reconcile what today's science says is true with what a book thousands of years old says about reality? Yes, I'd discovered that both the NT and QP share an affinity for translogical thinking, but how far did that similarity go? How alike were the biblical and scientific worldviews, really?

Being a scientist, I sought an answer to this million-dollar question by using a systematic, evidence-based strategy. I determined where science and the Bible—Christianity, in particular—stood on major issues.

I don't have the space here to report everything I discovered, but my exploration included three popular Western worldviews: scientific, atheistic, and Christian. And it asked each of them three deep questions: (1) Does absolute truth exist, or is truth entirely relative? (2) Are there truths that cannot be proven? (3) Is the universe designed for life?

I'll present the results in a matrix, for ease of comparison:

	SCIENTIFIC	ATHEISTIC	CHRISTIAN
Does absolute truth exist?			
Are there truths that cannot be proven?			
Is the universe designed for life?			

When I first did this exercise, my comparison included *all* the religions I had studied. And I asked them more questions than these three—for example: What is the nature of time? Is there life after death? Are humans unique?

But I believe this small sampling will give you a good sense of what I found—and that its results will surprise you as much as they did me.

SCIENTIFIC WORLDVIEW

Does absolute truth exist? YES

In the late seventeenth century, Isaac Newton discovered evidence that the gravitational force exists throughout the entire universe, not just here on Earth. This absolutely shocking revelation contradicted the long-held Aristotelian belief that the universe was divided into two nearly opposite realms: the *terrestrial* and the *heavenly*.

Think of it as the cosmic equivalent of the separation of church and state.

The terrestrial realm—which is everything from the moon on down—was said to be corruptible and changing. It was made of and governed by the mixing and matching of four imperfect elements: earth, air, fire, and water.

The heavenly realm was supposedly incorruptible and unchanging. It was made of, and governed by, a single, perfect element: the *quintessence* (literally, the *fifth* element).

Understandably, then, people were stunned to learn that not just Earth but also the starry heavens were subject to gravity's secular influence; that gravity is not merely a regional force but a *universal* one described by a universal law—a law that's now taught in high school:

$$F = GMm/r2$$

To put it simply, there isn't one law of gravity for you and another one for me. There isn't one truth for you and one truth for me. *Your* law of gravity is identical to *my* law of gravity.

Whether you are rich or poor, Muslim or Christian, white or black, male or female, if you jump off a cliff, you will drop at thirty-two feet per second per second. It's an absolute truth.

One more thing. Science has identified certain superficial aspects of reality that are relative; they depend on your frame of reference or point of view. But even *they* are subject to laws of physics that are absolute.

Case in point: energy, matter, time, and space. They are relative quantities, as I'll explain later in the book. But they obey the strict, absolute, and universal laws of special and general relativity.

***Are there truths that cannot be proven?* YES**

Both science and mathematics agree on this one, so I'll give you a quick example of each one.

First, science. Consider Einstein's famous equation from his theory of special relativity: $E = mc^2$. We now have more than a century's worth of evidence that it's true.

Is that proof? No.

Why not? I'll let Einstein answer it for you: "*The truth* of a theory can never be proven. For one never knows if future experience will contradict its conclusion."[1]

Over the years, Einstein's cogent observation has been paraphrased thusly: "No amount of experimentation can ever prove me right. A single experiment can prove me wrong."

So, then, even though special relativity appears to be true, science can't ever *prove* it.

Second, mathematics. Say hello to Kurt Gödel. After Aristotle, he's the greatest logician who ever lived.

In 1931, the young Austrian proved what's now called Gödel's incompleteness theorem (GIT).[2] The mathematics of it is a bit complicated, but its central message is not.

According to GIT, there are *many* truths that can never be proved using conventional logic. Yes, you read that right. Ordinary Aristotelian logic simply isn't powerful enough to affirm the veracity of certain big, complicated truths.

I like to summarize Gödel's astonishing findings this way: *Truth is bigger than proof.*

Here's yet another way to think about it. What happens when you ask too much of your computer? For example, when you open too many windows at once. The computer crashes.

GIT proves—*proves*—that logic is like that.

Ordinary logic is quite good at proving simple truths, such as the Pythagorean theorem and all those other elementary theorems you proved in geometry class. Remember that? But logic crashes when you ask it to prove big, complex truths.

Consider, for example, this big, bold statement from the Declaration of Independence: "All men are created equal." Is it true? Not true? The question cannot be settled using logic. Believing this "self-evident" truth requires faith.

I call statements like this—that are absolutely true but can never be *proved* true—*Gödelian truths* or *translogical truths*.[3] They defy—*transcend*—the pedestrian rules of ordinary reasoning.

Is the universe designed for life? YES

Meet cosmologist and astrophysicist Sir Martin Rees, Britain's Astronomer Royal. He's an Atheist who once declared, "I've got no religious beliefs at all."[4]

Sir Martin wrote a positively wonderful book titled *Just Six Numbers*. In it, he itemizes what I call *six vital signs* of the universe.

When you go in for a checkup, the doctor measures your vital signs—temperature, blood pressure, pulse, breathing rate, and so forth—to see how you're doing. Sir Martin's six numbers tell us how the universe is doing. And when we take measure of them, the results are shocking.

The six vital signs encompass exactly—not approximately, *exactly*—the values necessary for life to exist in the universe. And not just human life, mind you, but any and all organic life forms.

"We have a lot of really, really strange coincidences," says Stanford physicist Andrei Linde, "and all of these coincidences are such that they make life possible."[5]

If any of the vital signs were off by even a smidgen, neither you nor I nor any of Earth's plants and animals—nor any life forms possibly existing on other worlds—would or could exist. The universe would be thoroughly desolate, like some cosmic-sized ghost town.

Is it really just coincidence? Did our universe simply get lucky?

Well, you could easily shrug off a few perfectly calibrated vital signs. But six or more?[6]

To comprehend the enormous improbability of our situation, take a close look at one of Sir Martin's vital signs: lambda (Λ), the cosmological constant. Lambda tells us something about how fast the universe is accelerating outward, like a gigantic inflating balloon of spacetime—an expansion we believe might be driven by a repulsive, omnipresent, invisible fog called dark energy. Lambda also tells us something about the age of the universe and about whether life in the universe is possible or not.

Your chances of winning the Powerball or Mega Millions lottery are about one in 175 million.[7] The chance of lambda having precisely the value necessary for life—which it does—is one in a trillion trillion trillion trillion trillion trillion trillion trillion trillion trillion. In scientific notation, that's $1{:}10^{120}$.

One way to explain away our astronomical good fortune is by assuming that many other universes exist. If it's true, then our good fortune isn't miraculous. Given enough contestants in a lottery, someone is bound to win—which happens often enough with Powerball and Mega Millions to keep people playing.

One particularly ardent supporter of the so-called multiverse theory is Sean Carroll, a physicist at Caltech. He says, "The number [of universes] could well be infinity. So it is possible that somewhere else in this larger structure that we call the multiverse there are people like us . . . thinking about similar questions."[8]

There's another way of explaining away our astronomical good fortune. Maybe we're overstating the importance of perfectly tuned vital signs. Maybe a universe with "bad" vital signs can have life. Yes, they'd need to be exotic life-forms that science cannot presently imagine, but it wouldn't be the first time science was taken by surprise.

I enjoy such theoretical musings and believe they're worth bearing in mind. But right now, there's no evidence for any of these speculations—and worse, no prospect of ever obtaining any such evidence.

For instance, after decades of trying, no one has been able to come up with a workable experiment that can detect the existence of multiple universes, and it's not very likely anyone ever will. After all, how can we hope to observe legions of universes out there somewhere over the rainbow when we can't even fully observe our own universe?

Lee Smolin, a physicist at Canada's fabled Perimeter Institute for Theoretical Physics, summarizes the dilemma this way: "The multiverse theory has difficulty making any firm predictions and threatens to take us out of the realm of science. These other universes are unobservable."[9]

Sabine Hossenfelder, a theoretical physicist at the Frankfurt Institute for Advanced Studies, agrees wholeheartedly. "Without making contact to observation, a theory isn't useful to describe the natural world," she says. "[It's] not part of the natural sciences, and not physics."[10]

For the present time, therefore, the hypothetical multiverse remains as fictional as the Land of Oz. And as otherworldly as heaven and hell.

In the meantime, there's no denying our universe is miraculous, having won the lottery not once or twice but at least six times over. It is indisputable evidence, admits Atheist astronomer Sir Martin Rees, that "we are in a privileged subset of all possible universes."[11]

A privileged cosmos exquisitely tailor-made for us.

ATHEISTIC WORLDVIEW

Does absolute truth exist? **NO**

There are many denominations of Atheism, as there are of any religion. Here I interrogate an especially popular sect, *Post-truth Atheism*, which holds that truth is relative. Truth depends entirely on feelings and experiences, which are totally subjective.

The movie *Altered States* assigns this belief to its chief protagonist, a psychologist who specializes in altered states of consciousness. "The final truth of all things," the Atheist scientist declares, "is that there is no final Truth."[12]

This belief is on full display in a YouTube video that features a young woman named Rebecca having a spirited disagreement with a group of peaceful demonstrators in Los Angeles.[13] What she says at one point, with obvious sincerity, pretty much sums up the post-truth worldview:

> **Rebecca:** As a human being, we should pay attention to fear and not logic.
>
> **David:** Wait, wait. You said pay attention to fear and *not* logic?
>
> **Rebecca:** Yes.
>
> **David:** I should pay attention to emotions and not facts?
>
> **Rebecca:** Yes. . . . Emotions are the only thing that are real in this world.

Are there truths that cannot be proven? **NO**

Fundamental to all atheistic worldviews is the belief that something is true if and only if it can be *proven*. Thomas Edison, American inventor extraordinarie, said it this way:

> I cannot accept as final any theory which is not provable. The theories of the theologians cannot be proved. Proof, proof! That is what I always have been after; that is what my mind requires before it can accept a theory as fact.[14]

In other words, for Atheists, *truth equals proof.*

I am intimately familiar with this worldview. As a scientific monk, I sided with Edison. I believed only in things that could be proved logically. You claim God exists? *Prove it!* Otherwise, shut up.

Is the universe designed for life? **NO**

Consider the remarkable life and work of Atheist Steven Weinberg, an American physicist and Nobel laureate at the University of Texas at Austin. Weinberg wrote a beautiful book titled *The First Three Minutes*, which purports to describe the universe during the three minutes immediately following the big bang.

With a long, distinguished career under his belt, Weinberg has come to a striking conclusion. "The more the universe seems comprehensible," he says, "the more it also seems pointless."[15]

For Weinberg and other Atheists, the universe and everything in it—you, me, them—are the products of a series of spectacular accidents. So all of it is meaningless—including Weinberg's career, his achievements, and his opinion that the universe seems pointless.

CHRISTIAN WORLDVIEW

Does absolute truth exist? **YES**

According to the New Testament, Pontius Pilate, the first-century Roman governor of Judea, asked Jesus of Nazareth: "What is truth?"[16]

Elsewhere, Jesus said to his followers: "I am the way and the truth and the life. No one comes to the Father except through me. If you really know me, you will know my Father as well. From now on, you do know him and have seen him."[17]

Notice two things here.

First, according to the Christian worldview, Jesus is the human equivalent of God. "The Father and I are one," Jesus declares in John 10:30.

In math, equivalence is much stronger than mere equality. One plus one *equals* two. But the number three is *equivalent* to the number three. The two are identical.

We designate equivalence with a special symbol: ≡

Thus, according to the Christian worldview, Jesus ≡ God.

Second, within the Christian worldview, there isn't a God for you and another God for me; we all live under the authority of a single, universal God—just as we all live under the constraints of a single, universal gravitational force.

"This is what the Lord says," reports the prophet Isaiah: "I am the First and the Last; there is no other God."[18] The apostle Paul, in his first letter to his protégé Timothy, declares: "There is one God and one Mediator who can reconcile God and humanity—the man Christ Jesus."[19]

Are there truths that cannot be proven? YES

The New Testament recounts the experience of a skeptical disciple named Thomas. When the other disciples told him that the recently crucified Jesus had returned to life, Thomas wasn't buying it. "Unless I see the nail marks in his hands and put my finger where the nails were, and put my hand into his side, I will not believe."[20]

Thomas's motto in life is a common one today: *Seeing is believing.* He would have made a good candidate for governor of Missouri, the "Show-Me" State.

But then Jesus appeared to Thomas and gave him a faith-building invitation: "Put your finger here, and look at my hands. Put your hand into the wound in my side. Don't be faithless any longer. Believe!"[21]

Thomas obeyed and was quickly convinced of the reality of Jesus' resurrection. "My Lord and my God!" he cried out.[22]

Jesus' reply is quite telling: "You believe because you have seen me. Blessed are those who believe without seeing me."[23]

According to Christianity, there are truths—like Jesus' existence—that cannot be proved. They must be believed to be seen.

The skeptical Thomas *believed* enough in the possibility that the crucified Jesus was alive that he reached out and touched the wounds. He *believed* enough in something that sounded outrageous to determine the truth of the matter for himself. If he hadn't—if he'd simply walked away in a dismissive huff—he would have forever remained in the dark.

The same goes for you. Unless you're willing to believe that something *might* be true, you'll never bother to investigate and see for yourself whether it *is* true (or not). You'll remain in a state of confident ignorance.

Is the universe designed for life? **YES**

According to the Christian worldview, the universe is not an accident, and neither are you. All of it is the deliberate creation of a brilliant, loving, rational being—God.

The Bible makes this crystal clear by opening with these famous words: "In the beginning God created the heavens and the earth."[24] From then on, the narrator describes with specificity the unfolding masterpiece, culminating with the intentional formation of the first spiritually intelligent creatures, *Homo sapiens sapiens*, Adam and Eve.

COMPARING AND CONTRASTING

When you fill in our matrix with all of the foregoing information, here's what you get:

	SCIENTIFIC	ATHEISTIC	CHRISTIAN
Does absolute truth exist?	YES	NO	YES
Are there truths that cannot be proven?	YES	NO	YES
Is the universe designed for life?	YES	NO	YES

Do you notice anything interesting? Anything surprising? When I first did this analysis, I noticed two things right away that took me aback.

First, the atheistic worldview is fundamentally *opposed* to the scientific worldview.

Second, the Christian worldview is fundamentally *in line* with the scientific worldview.

This outcome shocked me because I'd always taken it for granted that Atheism was wholly in line with science, my life's pursuit. And that Christianity was wholly in conflict with my beloved science.

Today, many outspoken Atheists perpetuate that lie. They boast that science is on their side and that Christianity is an antiscientific, Bronze Age superstition. But as you can see, they're greatly mistaken. And so are the many Christians who see science as an enemy, as an institution out to undermine their Bible-believing worldview.

As a scientific monk and devout Atheist—an avowed free-thinking intellectual—I couldn't shrug off the results of this straightforward comparison. A comparison, remember, that included *all* the religions I had diligently explored and many more than just these three big questions.

The results rocked my lifelong secular worldview far more so than even the missing-mass problem had done. But I still wasn't ready to come to a definitive conclusion. I still had more questions that needed answering.

4

THE RECKONING: PART 2

It was the best of times, it was the worst of times,
it was the age of wisdom, it was the age of foolishness,
it was the epoch of belief, it was the epoch of incredulity,
it was the season of Light, it was the season of Darkness.

CHARLES DICKENS, *A TALE OF TWO CITIES*

As a grad student at Cornell, I was stunned to discover how well the scientific and Christian worldviews lined up on the fundamentals. After all, the two were widely perceived—including by me at the time—to be hoary antagonists.

I remained a practical Atheist who lived as if God didn't exist, whether he did or not. But now I was eager to find out how wide and how deep was this apparent science/Christianity accord.

Here are two more of the many deep questions I explored during my investigation: Are humans unique? Is light special?

These particular questions were especially important to me because science had discovered strong, remarkable answers to both. If the answers offered by Christianity differed from those firmly held by us scientists, it would be game over. I would ditch Christianity, just as I had all the other religions, and stick solely with my science and Atheism.

ARE HUMANS UNIQUE?

Christian Worldview YES

According to a common interpretation of the Bible, every vertebrate is a soul—*nephesh*, in ancient Hebrew. If so, it means that you, your pet dog, cat, or ferret each represent a living soul, a *nephesh chayyah*.[1]

According to mainstream Christianity, each and every human also has something else: a *spirit*. Every person is made in God's image, and the Bible describes God as a preternatural, spiritual being. "God created human beings in his own image. In the image of God he created them; male and female he created them."[2]

According to Christianity, therefore, you are a soulful, spiritual creature temporarily inhabiting a physical body. It's what makes you different from every other animal on the planet. It's what makes you unique.

Scientific Worldview YES

When I was a kid, my science teacher taught us that humans are unique because we alone have intelligence. But times have changed. Science now knows that lots of animals have IQ: whales, dolphins, dogs, pigs.

So, then, my old science teacher was wrong: IQ does not make us unique.

Are there any other traits that make humans unique?

At first blush, the answer isn't obvious. All too often, even today in the twenty-first century, we behave just like other animals—except we're all dressed up.

Humans can speak—but so do other animals. Some African grey parrots have vocabularies of up to a thousand words.[3]

Humans create impressive societies, invent clever tools, design soaring skyscrapers—but so do other animals. Bees live in sophisticated, highly organized societies; otters use rocks to hammer open clams;

chimps repurpose twigs to fish for termites. And speaking of termites, they're famous for designing and building eco-friendly and architecturally brilliant skyscraper-like mounds out of mud, dung, and saliva.

Humans have an impressive genome—but, again, so do other plants and animals. In fact, the largest known strand of DNA belongs to *Paris japonica*, a white, star-like flower native to Japan. Its DNA is fifty times larger than ours.

And the African marbled lungfish, *Protopterus aethiopicus*, has the animal world's largest genome. It's more than forty times bigger than ours.

Humans sit atop the food chain—but not even this makes us uniquely special. Have you ever played the game Jenga? There's a tower of interwoven wooden blocks and you have to remove them one by one without toppling the whole thing.

Obviously, if you remove the topmost block, nothing happens. That's the point. Remove humans and the environment won't care—except that the remaining plants and animals would probably cheer. Our position atop the food chain, in other words, makes us uniquely *in*significant.

So, then, scientifically speaking: Is there *anything* that makes humans unique? *Yes.* And it's not just one thing.

According to paleoanthropologists and other scientists, we *Homo sapiens sapiens*—anatomically and behaviorally modern humans—appeared on Earth suddenly, not gradually.[4] Prehistoric fossil and genetic records are ambiguous, but they indicate we happened onto the scene less than 100,000 years ago—like yesterday, given that science estimates Earth to be more than four billion years old.

What's more, we humans burst onto the stage bundled with a cluster of behaviors never before seen on Earth. Sarah Wurz, an archaeologist at the Institute for Human Evolution, University of the Witwatersrand, Johannesburg, South Africa, explains that these unique behaviors include "art, religious beliefs, and complex technologies," as well as "the ability to communicate habitually and effortlessly in symbols . . . [and] advanced problem solving and long range planning abilities."[5]

UCLA's Pulitzer Prize–winning evolutionary biologist Jared Diamond calls our abrupt and unique appearance on Earth *the great leap forward.* "Insofar as there was any single moment when we could be said to have become human," he says, "it was at the time of this Great Leap Forward 35,000 years ago."[6]

"What happened at that magic moment in evolution? What made it possible, and why was it so sudden?" he asks. "This is a puzzle whose solution is still unknown."[7]

SO WHAT?

As a young man teaching physics at Harvard, I was shocked by the compatibility of these scientific findings with the Christian worldview. It didn't mean to me that science *proved* the Christian worldview because science doesn't have the power to do such a thing. But I saw that science didn't *contradict* the Christian worldview—which is no small thing.

Besides teaching at Harvard, I authored a monthly column in *Psychology Today*, called "Rational Alternatives," in which I opined about the latest research in psychology (a "soft" science) from the perspective of a hard scientist.

Because of this, I was keenly aware that Harvard psychologist Howard Gardner was making headlines with a revolutionary theory about multiple intelligences. "The idea of multiple intelligences," he explained, "was developed to document the fact that human beings have very different kinds of intellectual strengths."[8]

In 1983, Gardner published his thesis in a landmark book titled *Frames of Mind: The Theory of Multiple Intelligences.* In it, he identifies no fewer than seven different intelligences. Two of these—visual/

spatial and logical/mathematical—form the basis of what traditionally we call IQ. The other five are bodily/kinesthetic, musical, interpersonal, intrapersonal, and verbal/linguistic.[9] Later, an eighth and ninth intelligence—naturalist intelligence and existential intelligence—were added to the list.[10]

A dozen years later—shortly after I left Harvard to work full-time at ABC News—my friend and fellow reporter Daniel Goleman introduced the concept of yet another intelligence: *emotional intelligence*, or EQ. Goleman even wrote an international bestseller on the subject, titled *Emotional Intelligence: Why It Can Matter More Than IQ*.

As I contemplated both Gardner's and Goleman's claims, I thought it odd that no one had yet talked about a *spiritual intelligence*—or SQ, as I came to call it. After all, one of the unique traits of our species is our spirituality: our religious art, literature, and music; our belief in supernatural deities and an afterlife; our habit of burying our dead with great ceremony; and our powerful religious passions, which routinely erupt in ways both heavenly and hellish.

To put it plainly: Spiritual intelligence is what makes us humans unique; no other animal on the planet has it. That is, for you and me, SQ is a nonzero positive integer. For all other creatures, it is zero.

Tangible evidence of our SQ is everywhere. In paintings such as *The Last Supper*; buildings such as the Notre-Dame Cathedral; literature such as *Paradise Lost*; oratorios such as Handel's *Messiah*; and countless other divinely inspired creations.

No other living being on Earth does such things.

IS LIGHT SPECIAL?

Christian Worldview YES

According to the New Testament, "God is light; in him there is no darkness at all."[11] The verse does not equivocate. It doesn't say God is kind of like light. Or that God is analogous to light. It states plainly and boldly: God *is* light.

It's like a mathematical equivalence:

$$\text{GOD} \equiv \text{LIGHT}$$

This helps define the God of the Bible. And it also gives light a *sacred* status.

Scientific Worldview YES

Everything in the universe behaves like either a *particle* or a *wave*. They are opposite kinds of reality, like odd and even numbers in mathematics. There is no in-between.

A rock, a car, your body all behave like particles. A particle holds its shape and travels in only one direction at a time.

A ripple, a tsunami, a sound blast all behave like waves. A wave smears out and travels in many directions at once.

That's what scientists always believed.

During the seventeenth century, for instance, Isaac Newton argued that rays of light behave just like particles. Different shapes produce different colors. Simple!

But Christiaan Huygens, a Dutch scientist, strongly disagreed. He claimed that rays of light behave like waves. Different wavelengths produce different colors. Also simple!

Who was right?

Fast forward more than a century to 1801, when British scientist Thomas Young did a very clever experiment to settle the argument. He made rays of sunlight squeak through two narrow slits and then studied the pattern created on a screen downstream.

Before Young did the experiment, he made these clear-cut predictions:

- If Newton is right, light particles will zip through the two slits and create two slivers of light on the screen.
- If Huygens is right, light waves will squeeze through the two slits and will alternately reinforce and cancel themselves out, creating a pattern of light and dark bands on the screen.[12]

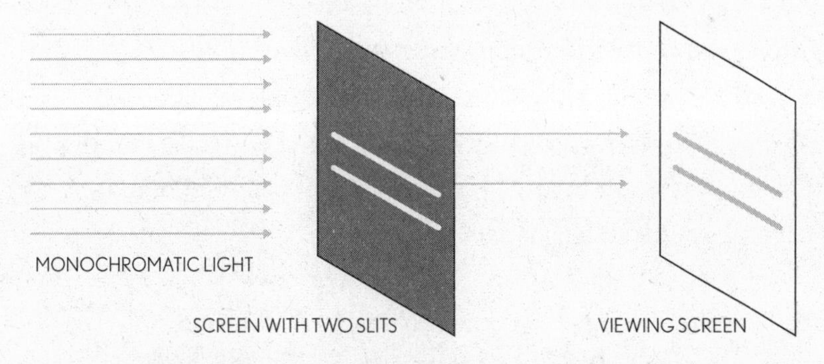

NEWTON: LIGHT BEHAVES LIKE PARTICLES

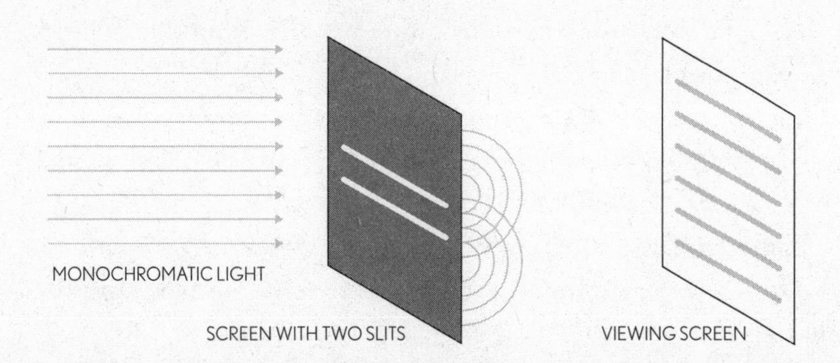

HUYGENS: LIGHT BEHAVES LIKE WAVES

So what did Young discover? His experiment produced alternating bands of light and dark! The verdict was unmistakable: Huygens was right. Light behaves like *waves*.

But that wasn't the end of the story.

Scientists went on to do still more snooping, and—much to their astonishment—they discovered five new things about light's true colors.

LIGHT'S TRUE COLORS

1. Light is the embodiment of a contradiction.

Eighty-six years after Young's bombshell experiment, a German scientist named Heinrich Hertz—no relation to the rental car company—did a different experiment.[13] He pointed a beam of light at a sheet of metal and studied what happened.

Lo and behold, Hertz discovered that light behaves like *particles*—directly contradicting what Thomas Young had found.

What in the world was going on?

Science was now in a real pickle.

Enter a scruffy-looking scientific nobody named Albert Einstein. In 1905, he won instant fame by proposing a Solomon-like solution to the problem.

Light, said Einstein, is something never before imagined in the universe: It is both particle *and* wave. Not a mash-up of half particle and half wave but *fully* particle and *fully* wave.

Einstein named this unimaginable chimera a *light quantum*. Today it's more commonly called a light photon or simply a photon.

Einstein's solution was like claiming that a number could be even *and* odd; a color could be black *and* white; a statement could be true *and* false. Yet there was no denying the evidence: Both Young's and Hertz's experiments were unimpeachable; their results, though contradictory, had to be believed.

Einstein's crazy idea ended up revamping the classical scientific worldview and producing quantum physics. For that monumental achievement, the wiry-haired genius was awarded the 1921 Nobel Prize in Physics.[14]

2. Light does not obey the rules of ordinary matter.

In grade school you learned to create any and all colors of paint by mixing together just three primary ones: red, blue, and yellow. Blue and yellow produce green; blue and red produce purple; red and yellow produce orange, and so forth.

What if you mix equal amounts of all three paint colors? You get black.

What you didn't learn in grade school is that paint is made of colored particles of matter. But light, being a weird wave-particle, behaves differently. *Very* differently.

For starters, the three primary colors of light are not red, blue, and yellow. They're red, green, and blue (RGB), from which the images on a color television and color computer screen are created. From RGB you can create the entire rainbow.

Also, if you mix equal amounts of the three primary colors of light, you don't get black. You get *white*! The exact opposite of mixing paints.

3. Light can transform into matter, and vice versa.

Light behaves very unlike matter, but it can become matter under certain conditions—most commonly by *colliding* with matter. For instance, if a bright-enough photon of light collides with a heavy nucleus, it spontaneously becomes an electron and its antimatter twin, a positron. We call this *particle creation*.[15]

The opposite is also true. Matter can change into light. An electron colliding with a positron, for instance, will explode into a pair of light photons. We call this *particle annihilation*.[16]

Either way, the transmogrification always happens in accordance with Einstein's iconic $E = mc^2$ equation, which describes the mysterious interchangeability of energy and matter.

4. Light has a sacred status in the universe.

According to special relativity, light enjoys a divine-like status in the universe. Light alone can travel at 186,000 miles per second. Nothing else—not you, me, or any material object—can ever reach that phenomenal speed, no matter how hard we might try.

5. Light exists in a timeless world.

Special relativity also says that *time is elastic*. As you speed up—say, aboard Elon Musk's Starship headed for Mars—your sense of time automatically slows. Your heart rate slows. Your aging slows.

This raises a very interesting question. Hypothetically speaking, what if you achieved the impossible and reached the speed of light? What would happen to your sense of time?

Answer: It would slow to a dead stop, at which point you'd become a timeless being. Time would cease to exist. There'd be no past, no future, only an eternal *now*.

SHINING LIGHT ON THE SCIENTIFIC AND CHRISTIAN WORLDVIEWS

When I was done with this exploration of light from a biblical and quantum mechanical perspective, I was shocked anew. For here was further evidence of what I'd already discovered—that the scientific and Christian worldviews are fundamentally compatible.

Here's what I mean.

1. Light is the embodiment of a contradiction.

The scientific and Christian worldviews agree that jarring contradictions can coexist comfortably. When they do, they represent profound truths—as I explained in chapter 2. Such possibilities, therefore, must be taken seriously and not summarily dismissed.

Here are two prime examples of meaningful contradictions:

- **Science:** Light is both a *particle* and a *wave*. In quantum physics, it's called *wave-particle duality*.
- **Christianity:** Jesus is both *mortal* and *immortal*.[17] In Christology, it's called the *hypostatic union*.

2. Light does not obey the rules of ordinary matter.

The scientific and Christian worldviews agree that it is impossible to get black by combining any colors of light. In particular, here's what they say:

- **Science:** When you combine all the colors of light, you get white.
- **Christianity:** "God is light; in him there is no darkness at all."[18]

3. Light can transform into matter, and vice versa.

The scientific and Christian worldviews agree that light can become matter, and vice versa. In particular, here's what they say:

- **Science:** In nuclear physics, these extraordinary transformations are called *particle creation* and *particle annihilation*.
- **Christianity:** The New Testament reports that God—who is light—became matter . . . a man in the person of Jesus. In Christianity, this extraordinary transformation is called the *Incarnation*.

> The Word became flesh and made his dwelling among us.
> We have seen his glory, the glory of the one and only Son,
> who came from the Father, full of grace and truth.[19]

Also, according to Christianity, following your material life on Earth, you are transformed into a spiritual being that returns to

God, who is light. This transformation is called *resurrection from the dead.*

> This is how it will be when the dead are raised to life. When the body is buried, it is mortal; when raised, it will be immortal.[20]

> He [Jesus] will take our weak mortal bodies and change them into glorious bodies like his own, using the same power with which he will bring everything under his control.[21]

4. Light has a sacred status in the universe.

The scientific and Christian worldviews agree that light—and its equivalent, God—has a sacred status in the universe. In particular, here's what they say:

- **Science:** According to special relativity, the speed of light—designated by the letter *c*—is a sacred number and a universal constant.
- **Christianity:** According to the Bible, God—designated by the Greek letters *alpha* and *omega*—is sacred and universal. "I am the Lord, and there is no other; apart from me there is no God."[22]

5. Light exists in a timeless world.

The scientific and Christian worldviews agree that light—and its equivalent, God—exists in a timeless world. Not eternal, mind you—timeless. Scientifically, there's a huge difference.

Eternal means you're stuck within a timeline that goes on forever and ever. *Timeless* means you exist outside of the timeline, outside of time.

In particular, here's what they say:

- **Science:** At the speed of light, time stops. Light, therefore, is not constrained by time. It is a timeless phenomenon.
- **Christianity:** God created time; so he exists outside of it. God, therefore, is not constrained by time. He is a timeless entity.

When I first read the biblical story about the burning bush, I found it odd that God says to Moses, "I AM WHO I AM. Say this to the people of Israel: I AM has sent me to you."[23]

But following my analysis of light from a scientific and biblical perspective, I realized it's exactly what a timeless being would say about himself. He would say: "I don't have a past, present, or future. I just am. *I am who I am.*"

One other thing crossed my mind: As I stated earlier, death is comparable to particle annihilation. When you die, you transform into photons of spiritual light that return to your Maker.

If that's true—and such a belief is entirely consistent with both the Bible and science—you necessarily escape the timeline. You enter a timeless realm as a spiritual being that just *is*—just like God himself. And you remain that way—*you are who you are*—forever.

THE RECKONING

As a scientist, I'm trained to gather data carefully and then to seek the most reasonable, most elegant, most streamlined explanation consistent with that data.

By 1995, I realized that after nearly two decades of gathering data—of passionately seeking out scientific and metaphysical perspectives on my original question: *How did this amazing, mostly invisible universe of ours come to be so amazing and mostly invisible?*—I was ready to sit down and reason my way to an optimal conclusion.

I knew it could be only a *tentative* conclusion. As a scientist, I knew that anything involving science is always tentative. Science—lacking the power to prove things, to settle matters once and for all—must always keep asking questions and seeking better and better answers.

But I knew I'd reached a significant point in my lifelong search for answers. It was now time to make some decisions.

5

THE RESULT

What then shall I do? . . . I will sing with my spirit,
but I will also sing with my mind.
1 CORINTHIANS 14:15, BSB

In late 1995, after some twenty years of studying science and exploring the world's religions, after comparing and contrasting their perspectives on virtually every important subject I could think of, I finally landed on a decision. I'd learned important and helpful lessons from each religion and deeply respected them all. But I chose the Christian worldview for at least two major reasons that are still true today.

One, the Christian worldview best answers my questions. Not *all* my questions—no worldview can do that, not even science.

Two, the Christian worldview best squares with the scientific worldview. It's easy for me to be both a scientist and a Christian.

Do science and Christianity have disagreements? You bet—and a few of them get the bulk of the publicity. But when it comes to the fundamentals, the two worldviews are very much in line.

They are like my wife and me. We have our disagreements, and some of them are real doozies. But when it comes to core principles, we see eye-to-eye.

MY COMING-OUT PARTY

After reaching my conclusion, I kept it to myself.

Why?

First, even though I was a public figure by then, I was still a scientific monk at heart. I still coveted my privacy.

Second, as a scientist and journalist, I regarded objectivity as something sacred. I worked hard to maintain a firewall between my public reporting and my personal beliefs. (By the way, I still feel that way.)

But then came a morning in 1997 that I'll never forget. I was sitting at a table inside the ABC News Times Square studio, with a small group of experts from various disciplines, doing a live segment for *Good Morning America* (*GMA*). We were discussing the ethics of cloning.

Months earlier, British scientist Sir Ian Wilmut had stunned the world by accomplishing what science had always said was impossible: He'd cloned a mammal, a sheep named Dolly, using a technique called somatic cell nuclear transfer. She was named after the country singer Dolly Parton and instantly became just as famous.

I was the first American television correspondent to interview Sir Ian Wilmut, and afterward I followed the story closely. It was very big news.

Day after day, it captured the attention of every major news organization in the world—from the AP and BBC to Kyodo News and Agence France-Presse. Even the august National Academy of Sciences, American science's equivalent of the Vatican, got into the act, convening official public forums to discuss the science and ethics of what was going on.

In the midst of this media frenzy, I managed to infiltrate various underground groups that were vowing to clone a human being, including a particularly bizarre cult called the Raelians. If any of these groups had succeeded, it would have been scientifically, ethically, and culturally historic.

As the *GMA* segment wound down—as the countdown clock came close to a *hard out*, a computer-driven exit to commercial—Charlie Gibson, the show's cohost, asked me for my final thoughts. Hastily, I said something to the effect of "Well, Charlie, I'm concerned that Wilmut's cloning technique might one day be used to clone a human being. It worries me not just as a scientist but as a scientist who happens to believe in God."

Instantly I froze. *What had I just said on national television?*

I was sure that viewers would be calling ABC News to complain. I'd lose the fan base I'd worked so hard to build. My bosses would not be happy that their number-one science correspondent had just confessed to believing in God.

In a fog, I rose from my chair and walked across the studio. But as I did, an amazing thing happened.

Stagehands, cameramen, and security guards stepped forward to shake my hand. They'd watched me on television for years and assumed I was an Atheist—*just because I was a scientist.* They were flabbergasted—thrilled—to learn that I believed in God.

When I got back to my office, my producer, Melissa Dunst, told me the switchboard was being flooded with phone calls about the cloning segment. I held my breath. *Here it comes*, I thought, *the beginning of the end.*

But it wasn't that at all. Melissa told me that nearly all the calls were incredibly positive. Viewers were elated that I believed in God. In fact, she said, I was even getting marriage proposals!

Five years later, I left ABC News to do other things, including becoming a father. I had long since decided that being a jet-setting correspondent, while exciting, was no life for a family man.

One of the first things I did was write a book in which I formally introduced my concept of spiritual intelligence (SQ). *Can a Smart Person Believe in God?* was my first attempt to put into words what I had discovered about science, Atheism, and Christianity. At the back of the book, I included the first-ever SQ test.

Among other things, I explained that SQ is a cognitive superpower unique to *Homo sapiens sapiens*. It allows you and me to sense, however imperfectly, Gödelian truths and translogical realities that cannot be seen, proven, or even imagined.

At the time the book was published, in 2004, medical science was just catching on to the importance of human spirituality and documenting its very real benefits to our well-being—a powerful repudiation of Atheists who mock religion.

I explained that the published research shows a high-SQ lifestyle is associated with lower risks of hypertension, heart disease, cancer, stroke, depression, suicide, drug abuse, and criminal delinquency. I wrote:

> The largest study to date tracked the lives and deaths of 21,204 adults for a full decade. The average results? A person who attended church at least once a week lived seven years longer than someone who didn't attend at all. Among African-Americans, the disparity was even more stunning: *fourteen years!*[1]

Today, similar results continue pouring in, so there's no doubt: A high-SQ lifestyle is extremely good for your health—as good or better than a nutritious diet and regular physical exercise.

STEREOSCOPIC VISION

In *Smart Person*, I explain SQ in terms of a 3D movie. Remember trying to watch one without those old cardboard red-and-blue glasses? All the images on the screen looked out of focus, right?

Figuratively speaking, that's how the universe looks to every nonhuman, spiritless animal on Earth: blurry and confusing. The world has no depth; what you see is what you get.

Bambi never looked up at the night sky and wondered if there was a God, nor did he blame him for his mom's senseless murder. He just saw stars.

This is also true if you look at the universe with your IQ or SQ alone. The universe—life—doesn't make sense. Everything is a bit confusing and lacks depth, lacks meaning, to the point of being depressing, or worse.

Your IQ and SQ are meant to work together, synergistically, like the lenses of those 3D glasses. Think of your IQ as the red lens and your SQ as the blue lens.

When you look at the universe through the two lenses, IQ *and* SQ, it no longer seems blurry or flat. You perceive depth and meaning. You see the universe in its full physical *and* spiritual grandeur.

You discern Gödelian truths and translogical realities more fascinating than even dark matter and dark energy; truths and realities that help you make sense of everyday life and the reason for your being.

According to Christianity, it is a foretaste of the privileged perspective you'll acquire after you die, when your immortal spirit rejoins its Maker. At that time, you'll behold not only the entire universe you left behind but the whole of reality. You'll be able to take it all in—reality's full, timeless grandeur—because your spirit will "see" it through the eyes of a God who is omniscient.

The New Testament describes the shocking transformation this way:

> For the present we see things as if in a mirror, and are puzzled; but then we shall see them face to face. For the present the knowledge I gain is imperfect; but then I shall know fully, even as I am fully known.[2]

According to a major survey by Reuters/Ipsos, 82 percent of the world's population believes that God (or gods) or some other Supreme Being (or beings) definitely or possibly exists. Only 18 percent say that such metaphysical beings do not exist.[3]

That's clear-cut evidence of our SQ, of our species' uniqueness. It's clear-cut evidence of our spiritual awareness.

Atheists routinely ridicule spiritual awareness; but as you can see, they are in the minority. They are, I submit, in denial. More about that later.

One of Atheism's favorite arguments is the *god of the gaps theory* (GOGT). It claims that any explanation that involves God or other supernatural beings is superstitious and no match for a bona fide *scientific* explanation. As scientific knowledge increases, Atheists claim, the need to invoke God or a Supreme Being will decrease.

GOGT's roots can be traced back to a nineteenth-century secularism popularized by positivist philosophers such as Auguste Comte, father of modern sociology. Positivists don't just love science, as I do; they *worship* it, as I once did.

For positivists, science is a religion and humanity's salvation. They believe that, given enough time, science will trump all other religions by explaining away all mysteries.

One such believer was Anthony F. C. Wallace, a Canadian-American anthropologist. In his enormously influential book titled *Religion: An Anthropological View*, he makes this bold prediction:

> The evolutionary future of religion is extinction. Belief in supernatural powers is doomed to die out, all over the world, as a result of the increasing adequacy and diffusion of scientific knowledge.[4]

If you agree with Wallace, with GOGT, you're ignoring what science has discovered about our species, which I explained in the previous chapter. You're claiming that our religious sensibilities are a *flaw*, rather than a unique, spectacular, defining characteristic of *Homo sapiens sapiens*.

But the evidence doesn't support Wallace's prediction or GOGT.

First, religious affiliation is *increasing*, not decreasing, as scientific knowledge soars. A Pew Research Center study titled *The Changing Global Religious Landscape* states it plainly: "The global share of religiously unaffiliated people is actually expected to fall."[5] Based on current world trends "people with no religion will make up about 13 percent of the world's population in 2060, down from roughly 16 percent as of 2015."[6]

Second, as scientific knowledge soars, the universe is becoming *more* mysterious, not less. Nowhere is this mystifying trend more on display than in modern physics.

Quantum mechanics, special relativity, general relativity—the theoretical pillars of modern physics—verge on the supernatural. Their ideas about quarks, gluons, the quantum vacuum, virtual particles, quantum entanglement, dark matter, dark energy, curved spacetime, multiverses, ten-dimensional strings, gravity waves, black holes, white holes, wormholes—and on and on—are no less otherworldly than any religion's God, gods, Supreme Being, or beings.

Consequently, over the past century, modern physics hasn't demystified the world—not by a long shot. Instead, it has increased by orders of magnitude our awareness of the universe's deep, jaw-dropping mysteries.

Even some honest-minded Atheists admit it. Here's neuroscientist Sam Harris:

> I don't know if our universe is, as J. B. S. Haldane said, "not only stranger than we suppose, but stranger than we can suppose." But I am sure that it is stranger than we, as "atheists," tend to represent while advocating atheism. As "atheists" we give others, and even ourselves, the sense that we are well on our way toward purging the universe of mystery. . . . Indeed, there are good reasons to believe that mystery is ineradicable from our circumstance, because however much we know, it seems like there will always be brute facts that we cannot account for.[7]

GOGT might still prove to be true, but I wouldn't bet on it. Certainly I wouldn't bet my life on it.

As you've seen—and will continue to see in subsequent chapters—the message is unambiguous. Our religious awareness, our SQ, is not a flaw. God is not a delusion.

Evaluated with an open, honest mind—and with those 3D glasses firmly in place—the latest scientific evidence is entirely consistent with believing there's far more to life and the universe than meets the IQ. And with believing that the God of the Bible is real.

END OF THE BEGINNING

In September 2000—just three years after I confessed my belief in God on national television—my fledgling scientific-Christian worldview was put to a severe test. I didn't see it coming.

It started with my receiving an invitation to visit the *Titanic*. Yes, *the Titanic*. The supposedly unsinkable ship whose decaying iron hulk rests at the bottom of the North Atlantic Ocean.

Honestly, I had mixed feelings about the extraordinary invitation. I was excited by it, of course, but also petrified because I had a deathly fear of water.

I trace the hydrophobia back to my parents, God bless them. For starters, Mom hated the water, and I picked up on it as a kid.

Also, I had a terrifying experience at the beach one day when I was no more than five. Dad, with me in his arms, and my uncle Marte—both great swimmers—waded into the deep water, where they proceeded to toss me back and forth like a football. It was all in good fun, and I'm certain I was safe in their hands, but I came away thoroughly spooked by the ocean.

What, then, should I do about the Titanic *invitation?* After hesitating, I accepted it—for two reasons.

First, having the chance to become the first reporter ever to visit the *Titanic* was the chance of a lifetime; I couldn't possibly wimp out.

Second, I now had a reputation on television for danger and disaster. Here are a pair of examples of what I mean.

On October 17, 1989, a magnitude 6.9 earthquake devastated San Francisco, causing 67 deaths and $5 billion in damage. I was working at *Good Morning America* at the time, and like scores of other reporters from around the world, I immediately flew to the scene of the disaster to cover it.

When I arrived, a senior geologist from the California Governor's Office of Emergency Services tipped me off to some big news. (One advantage I've always had as a journalist is that fellow scientists relate to and trust me more than they do generic reporters; so I routinely scoop the competition.) Her team, she told me, had pinpointed the quake's epicenter in the rugged Santa Cruz mountains, about sixty miles south of San Francisco. It was reportedly marked by a gargantuan surface crack.

That evening, I sallied forth with the geologist and my cameraman—over the objections, I might add, of my LA-based producer, who considered it too dangerous. For hours, we drove in utter darkness past police checkpoints and landslides that littered the winding roads with huge boulders.

Finally, in the wee hours of the night, we found it: a gaping wound in the ground wide enough for us to scramble down into. The night was pitch black, so we had to film the enormous crack by the light of our SUV's headlamps.

After wrapping, I hustled back to San Francisco, dropped off the videotape, and beat it to my hotel room to shower and shave for the start of the show. It was a close call, but the footage of the epicenter and my commentary proved to be so dramatic that *GMA* led *both* its hours with my report. That very rarely happened.

In early 1991, I flew to Kuwait City to cover the Persian Gulf war. Just before departing, I watched a *60 Minutes* report featuring the science advisor to Syria's president and Carl Sagan, my old professor from Cornell.

Both men were predicting that the oil-well fires set by Iraqi soldiers in Kuwait would have apocalyptic consequences. The smoke, they said, was rising high into the air and would soon begin blocking sunlight, plunging Earth into a nuclear winter, a calamitous season of planetary cooling like the one we believe wiped out the dinosaurs sixty-six million years ago.

Since my years with Carl Sagan at Cornell, he had become not only famous but also hyperpolitical. In particular, he'd been lobbying hard, though unsuccessfully, for the US to denuclearize, claiming that an all-out, global nuclear war would cause a nuclear winter.

I immediately recognized that Sagan's claim about the Kuwaiti oil-well fires was political, not scientific. I knew it because stratospheric winds are the only ones powerful and global enough to make any kind of pollution spread planetwide. And I had learned from scientists on the ground in Kuwait that the smoke shooting into the air from the oil-well fires did *not* have anywhere near a large-enough injection velocity—that is, sufficient upward thrust—to reach the stratosphere, which at midlatitudes begins roughly six miles above the earth's surface.

As soon as I arrived in Kuwait City, therefore, I went on the air to set the record straight. At worst, I explained to our viewers, the smoke would be a regional problem and quickly dissipate.

But that was just for starters.

Days later, my producer befriended a young Aramco engineer who offered to escort us safely past the myriad landmines protecting the burning oil fields. That night, being careful not to tip off the other reporters, our convoy set off. For hours, we held our collective breath while threading our way past countless booby traps. Finally, as the sun rose above the vast desert landscape, we arrived at the oil fires.

They resembled gigantic, vertical Roman Candles that were roaring like jet engines. Even from a distance you could feel their searing heat; I felt as if I were standing in front of a powerful blast furnace.

I wanted to get even closer for my standup—my short speech to the camera. But my South African cameraman balked at the idea, fearing that the heat would fry his expensive new equipment.

After some lively back-and-forth, I made him a proposal. "I'll memorize my lines," I said, "and at the count of three, we'll both run in, shoot the standup *on the first take*, then beat it out of there, okay?"

He agreed. And that's how I became the first correspondent to report from the hellish oil-well fires.

Shortly after the end of the Kuwaiti portion of the war, Red Adair, the famous Texas oil well firefighter, helped extinguish all six-hundred blazing conflagrations in a matter of months. As I had reported, there was no nuclear winter—and never any danger of one.

Given my reputation at ABC News for derring-do, the *Titanic* story seemed to fit right in. It's what my viewers had come to expect from me.

But this time, something very *un*expected was about to happen.

MY *TITANIC* EXPERIENCE

The adventure began when my production team and I flew to Halifax, Nova Scotia, to meet up with the captain and crew of the *Akademik Mstislav Keldysh*, a 6,240-ton Russian research vessel. The *Keldysh*'s pair of small, manned, deep-water submersibles, *Mir 1* and *Mir 2*, were capable of going all the way to the bottom of the Atlantic Ocean.[8]

The *Keldysh* set off and a little more than a day later arrived at our destination. The captain announced that we were now at exactly the spot—four hundred miles southeast of Newfoundland—where the *Titanic* had collided with an iceberg, broken in two, and sunk.

That night, I stood at the railing of the ship's top deck and stared up at the bright, flickering stars above. Then I lowered my gaze and looked out over the restless sea, which stretched before me for as far as the eye could see.

With my headset on, I listened to the *Titanic* movie soundtrack and tried imagining what it must've been like for those poor, doomed passengers. I realized that this ice-cold, menacing ocean was the very last thing on earth they saw before going under.

I fought hard to quell the hydrophobia swelling up inside me.

Days later, after attending many orientation sessions, my time was at hand. With my heart thumping like a bass drum, I clambered into one of the small submersibles, together with my diving buddy—the famous British comedy writer Brian Cooke—and our pilot, Viktor.

Inside the sub's cramped cabin, Brian and I lay stomach-down on small, padded benches and stared out through small, roughly nine-inch-diameter portholes. Viktor seated himself between us, facing a much larger porthole, above which was a control panel filled with all kinds of dials, switches, and lights.

Corkscrewing down at roughly one mile per hour, it took *Mir 1* about two-and-a-half hours to hit bottom. It's an experience I'll never forget. When Viktor switched on the sub's floodlights, my eyes beheld a vast bed of light-colored sand that looked like the lunar surface.

There isn't much sea life that can survive at that extreme depth. All I saw were titanium-white, slender, rat-tailed fish and titanium-white, delicate sea stars. The creatures lack color because there's no sunlight at the bottom of the ocean to provide it.

Shortly after *Mir 1* touched bottom, Viktor switched off the lights. Immediately, I felt us rise and glide forward. I pressed my face against the porthole, straining to see something, *anything*, in the pitch blackness.

Minutes later, a vast wall suddenly appeared right before my eyes. It was studded with giant—*what?* I couldn't figure it out. Then I knew. *Rivets!* I was staring at the riveted hull of the sunken ship.

"*Titanic*," intoned Viktor in his thick Russian accent.

It was one of the most chilling moments of my life. But the electric feelings of awe and disbelief quickly gave way to an overwhelming sense of sorrow.

As our tiny sub hovered before the ship's towering bow, Brian and I observed a moment of silence. I'm not ashamed to admit it: I wept as I thought about the scores of people who had drowned there.

During the next hour or so, Viktor took us on a grand tour of the wreckage. After exploring the ship's forward section, we flew across the debris field. Beneath me, I could see, half-embedded in the sand, ladies' shoes, leather valises, unopened crates of champagne—everything that had spilled out of the fractured, sinking ship.

As we approached the end of the debris field, I saw in the near distance the *Titanic*'s stern and one of its giant, surprisingly shiny propellers.[9] It seemed to me we were heading toward it too fast—and, worse, *accelerating*! Later, I learned that our sub accidentally got caught in a fast-moving, deep-underwater current.

A split-second later, *Mir 1* slammed into the *Titanic*'s propeller. I felt the shock of the collision; shards of reddish, rusty debris showered down on our submersible, obscuring my view through the porthole.

Brian and I exchanged anxious looks, as if to say, *What just happened? What's going on?* But neither of us said a word. Viktor was sitting up in his seat now, staring intently at the control panel. We didn't want to distract him.

I kept peering through my porthole, but there was very little I could see. I also kept glancing at Brian, who appeared to be growing restless.

"I told you I didn't want to go to the stern," he muttered. And it was true. For whatever reason, while we were preparing for our dive, Brian had told me that. But I'd forgotten until now.

My mind flashed back to a story the captain had told us during one of our orientation sessions. A man stuck inside a disabled sub had panicked. He'd lunged for the hatch to escape and opened it. Ocean water poured in, drowning him instantly.

As a precaution against Brian doing something similar, I repositioned my body on the bench, lowering my left foot onto the sub's floor. My thought was: *I'll tackle Brian if he so much as* looks *at the hatch of our sub.*

Ten minutes passed. Twenty. Thirty . . .

All the while, I could hear the engine straining to dislodge *Mir 1* by moving it forward and backward, forward and backward. Clearly, Viktor was trying to rock us out of our stuck position. Equally clearly, it wasn't working.

Also during this time, Viktor was communicating urgently with various persons via hydrophone. The tense dialog was in Russian, so I had no idea what was being said.

The voices on the hydrophone sounded wavy and echoey, as if they were coming from another world. The eeriness of it all and Viktor's somber demeanor added to the fearsomeness of our predicament.

A voice inside my head spoke up: "This is how it's going to end for you." Even now, as I write this story, I can hear those exact words.

I began recalling all the dangerous news stories I'd done. The countless earthquakes, volcanos, and hurricanes I'd survived. The mudslides, oil spills, and wars.

I thought back to the deadly, cold weather I'd experienced at the North and South Pole; to being apprehended by Egyptian soldiers in Cairo; to being stranded on the island of Cebu, in the Philippines.

I'd survived them all. But now . . .

"This is how it's going to end for you."

Suddenly, like a boot to the backside, my scientific worldview kicked in. I started brainstorming possible solutions. I'd always lived by the motto *For every problem, there is an optimal solution.*

But it wasn't easy coming up with solutions of any kind. I couldn't just call AAA to have us towed out.

I knew *Mir 2* was in the water; but I didn't know exactly where. Besides, even if it could get to us in time, how would it pull us out without endangering itself?

When I decided there was no escaping, I started calculating how much oxygen we probably had left. I figured we were good for another eight to ten hours at the most. Then we'd slowly suffocate.

That's when I thought of Laurel—and a heavy, crushing, depressing sadness fell upon me. I'd never see her again. *Never.* I couldn't believe it.

Then I thought of all the passengers who'd lost their lives on the *Titanic*. Soon, I'd be joining them and, like them, become a ghostly memory.

Then my fledgling Christian worldview took over. I wondered if it was really true that death was a rite of passage, a phase transition from one kind of existence to another. Like childbirth.

My scientific worldview broke in again.

Maybe the phase transition was akin to a melting ice cube, where water goes from a solid to a liquid while maintaining its chemical identity. Or maybe it was like particle annihilation, where matter instantaneously becomes energy. Or maybe it was like metamorphosis, where a caterpillar transforms into a butterfly.

Life after death, I mused, certainly wasn't far-fetched. There are many natural processes comparable to Christianity's concept of an afterlife.

Then something happened that's difficult to describe. The feel of the sub's interior space abruptly changed somehow. It was as if an invisible presence had entered it. At the same time, an uncanny and unheralded sensation of peace washed over me.

Shortly afterward, everything went quiet. *Mir 1*'s engine stopped roaring. Suddenly, it felt as if we were floating.

I traded looks with Brian, as if to say, *Could it be . . . ?* A moment later, I looked in Viktor's direction. "Okay? I queried.

Turning to me, Viktor flashed a big, fat smile and said just two words in his thick Russian accent: "No prro-blehem!"

Somehow, Viktor had managed to free *Mir 1* from the propeller. Later, I learned he was an experienced MiG pilot, accustomed to handling crises.

A few months after the incident, Laurel and I were reading the Bible when we came across this psalm:

> Where shall I go from your Spirit?
> Or where shall I flee from your presence?
> If I ascend to heaven, you are there!
> If I make my bed in Sheol, you are there!
> If I take the wings of the morning
> and *dwell in the uttermost parts of the sea*,
> even there your hand shall lead me,
> and your right hand shall hold me.[10]

As a conscientious scientist and journalist, I've always done my level best to accurately recount my experiences. This includes the one aboard *Mir 1* twenty long years ago. I don't claim to fully understand it, and I gladly leave it to you to decide for yourself.

This much I can say with absolute certainty: For as long as I live, Psalm 139 will never again be merely words in the Bible.

Down there in the uttermost parts of the North Atlantic Ocean, stuck inside a tiny Russian sub, I believe it's possible I *experienced* that psalm—experienced God's presence and peace, right when I was resigned to kissing my life good-bye.

GOT FAITH?

People like to throw around the word *faith*. Yet, when pressed, many of them can't explain what it is exactly. I've found this to be true for both my fellow scientists and fellow Christians.

Sincere, churchgoing Christians piously quote Scripture—chapter and verse—avowing that faith can move mountains, even faith as small as a mustard seed. Yet they struggle to explain faith beyond saying it's a supernatural kind of trust, as described in John 3:16: "For God so loved the world that he gave his one and only Son, that *whoever believes in him* shall not perish but have eternal life."[11]

Likewise, sincere, orthodox scientists talk about faith as if it were a four-letter word. To them, science is a rigorous, evidence-based discipline and faith is just the opposite: a loosey-goosey superstition embraced by weak-minded people.

To them, faith is always and only *blind* faith. That is:

FAITH ≡ BLIND FAITH

It seems that many scientists are unwilling or unable (or simply refuse) to see the enormous difference between *blind* faith and *evidence-based* faith. Or that the entire edifice of science itself is built on a foundation of both blind and evidence-based faith.[12]

My own interest in faith was piqued when Laurel and I first cracked open the Bible and read a verse reporting what Jesus said to a sick woman who was asking for his help: "Daughter, your faith has made you well. Go in peace. Your suffering is over."[13]

What!?

As a person trained to know all about natural forces, I was gobsmacked by this passage. Here was a clear-cut assertion that faith isn't merely a heartfelt belief in something seemingly far-fetched.

This verse (and many others, I quickly found out) claims that faith—far from being some benighted way of thinking or woo-woo-like, magical belief—is *an honest-to-goodness power of nature*, like electromagnetism and gravity. It's a power that is exceptionally important to God, who inhabits an unearthly realm; but when it's unleashed here on Earth, faith can effectuate measurable, physical, miracle-like changes in a person or situation.

Driven by my unquenchable curiosity, I decided to figure out what the phenomenon of faith is, exactly. To apply the full substance of my scientific training to study and explain in terms anyone can understand the inner workings of this profound, natural-preternatural ability of ours.

That was more than thirty years ago. In the following section, I'll report to you what I discovered. You'll see for yourself that evidence-based faith is profound, indispensable, and potent. In fact, it is the *mightiest power in the universe.*

THE TRUTH ABOUT FAITH

If you believe not, you shall not understand.

ST. CYRIL OF JERUSALEM

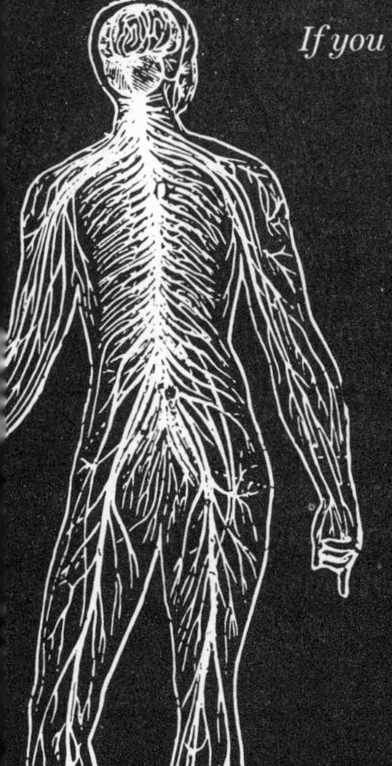

6

HAVING FAITH IN FAITH

Understanding is the reward of faith.
Therefore, do not seek to understand in order to believe,
but believe that you may understand.

AUGUSTINE OF HIPPO

I have two big questions for you.

First, what is faith, exactly? As a physicist, I'm looking for more than dictionary definitions, abstruse philosophical descriptions, or heartfelt personal testimonials. I want a rigorous, objective, concrete explanation consistent with the best available evidence.

Second, why in the world should you care what faith is? Why does it matter to your daily routine of getting up, going to work, dealing with people, paying the bills, keeping yourself and your friends happy, eating the right foods and watching your weight, reading the news and deciding what's true and what isn't?

Here are my answers, in reverse order.

First, you should care about faith because everything in your life depends on it. *Everything.* Your decisions, your relationships, your contentment: Every aspect of your being, right down to the functioning of your brain cells—as you'll see in a minute—depends on this thing called faith.

Second, as to what faith is, exactly, I will take you to its very source—like a guide leading a band of intrepid explorers to the headwaters of the Amazon, the mightiest river on Earth. You will behold for yourself the hidden, mind-boggling app that empowers you to believe in Gödelian truths and translogical realities—things that cannot be seen or proved or even imagined.

For starters, we must understand the difference between two main categories of faith: *misguided* and *enlightened.*

Atheist and journalist H. L. Mencken claimed that *all* faith is misguided. "Faith may be defined briefly as an illogical belief in the occurrence of the improbable," he writes. "A man full of faith is simply one who has lost (or never had) the capacity for clear and realistic thought. He is not a mere ass: he is actually ill."[1]

But Mencken was mistaken; not all faith is boneheaded. Here's how I explain it:

- **MISGUIDED FAITH** causes you to believe in known fantasies or hypotheses that can be conclusively discredited.
- **ENLIGHTENED FAITH** causes you to believe in hypotheses that are consistent with the best available evidence, even if that evidence is sketchy and ambiguous.

Moreover, each of these categories comes in two varieties: *IQ-based* and *SQ-based.* Here's what I mean:

- **IQ-BASED FAITH** causes you to believe in *trivial truths*—garden-variety, logical truths and realities that can be seen, proven, and imagined. Like this syllogism, for example:

  Any whole number divisible by two is even.
  Six is a whole number divisible by two.
  Therefore, six is an even number.

The argument is true but *trivial.*[2] There's nothing deep or mysterious about it. It requires only an IQ-based faith to trust it.

- **SQ-BASED FAITH** causes you to believe in *profound truths*—including Gödelian truths and translogical realities, which cannot be seen, proven, or imagined.[3]

Putting all this information into a matrix, here's what you get:

	MISGUIDED FAITH	ENLIGHTENED FAITH
IQ-BASED FAITH	All dogs have four legs. Fido is a dog. Therefore, Fido has four legs. *Dogs can be seen, proven, and imagined. But the argument's premise is false: not all dogs have four legs.*	The sky is blue. Earth is a planet.* *These simple realities can be seen, proven, and imagined. They're consistent with the best available evidence.*
SQ-BASED FAITH	The Tooth Fairy Easter Bunny *These cannot be seen, proven, or imagined. But they're also well-known fantasies created for kids.*	The Quantum Vacuum The God of the Bible *These cannot be seen, proven, or imagined. But they haven't been disproven, and both are consistent with the best available evidence.*

* This is true by definition

As I explained in the preface, what you choose to believe in life is all-important because your beliefs define your *worldview*. In times of trouble, they will spell the difference between your soaring or sinking. I'll have much more to say about this in chapter 12, "Having Faith in Your Worldview."

Take, for example, *The Crown*, a TV series that dramatizes the life and times of Queen Elizabeth II of Great Britain. In the episode titled "Moondust," her middle-aged husband, Prince Philip, announces he has lost his faith in God, in anything bigger than his own, aging self.

His new, atheistic worldview is depressing him, plunging him into a deep, existential crisis. It's a desolate worldview that he describes in terms of the 1969 Apollo 11 mission to the moon:

> The loneliness and emptiness and anticlimax of going all that way to the moon to find nothing but haunting desolation, ghostly silence, gloom. *That* is what faithlessness is. As opposed to finding wonder, ecstasy, the miracle of divine creation; God's design and purpose.[4]

What you *believe* dictates how you *see* life, others, and the world around you.

Believing is seeing.

How you see things, in turn, dictates how you react to circumstances, to crises, to everything.

Seeing is reacting.

The worst, most dangerous kind of worldview is powered by misguided faith. The best, most salubrious kind of worldview is powered by enlightened faith.

Where exactly do misguided and enlightened faith come from? Why are we wired for both? That's the deep mystery I'll tackle next.

THE SEAT OF FAITH

When I was at ABC News, I did a story on the Harvard Brain Bank, the world's largest repository of human gray matter.[5] I'll never forget cradling in my hands an actual human brain—the fragile, rubbery, three-pound vessel of a person's worldly essence. It was both thrilling and creepy.

The human brain, as I explained in my report, consists of two lobes that communicate via the *corpus callosum*, a thick, telephone-like cable of more than 300 million nerve fibers. Although the lobes are alike in some ways, they have different weights, shapes, and sizes. Even their functions are not entirely the same.[6]

I liken the lobes to fraternal twins. They're an odd couple of half-brains that are both competitive and complementary:

- **The left hemisphere** (LH) controls the body's *right* side and excels in logic, speech, and numerical calculations. Because it prizes detail, this twin prefers to think *analytically*, to break complexities down to their simplest elements.
- **The right hemisphere** (RH) controls the body's *left* side and has a precocity for intuition, sounds, and imagery. Because it values context, this twin favors thinking *holistically*, piecing together facts into a sensible, coherent big picture.

The LH focuses on individual trees; the RH takes in the entire forest. The LH relishes proof; the RH values hunches. The LH prefers to deconstruct; the RH loves to synthesize.

Decades ago, it was popular to claim that a person is either right- or left-brained. Scores of books pushing this trendy idea soared to the top of the *New York Times* bestseller list.

That notion has been thoroughly discredited. We now know that everyone is whole-brained, with a fully functioning LH and RH.

"The two hemispheres are both very competent at most things," explains UC Berkeley neuroscientist Richard Ivry. But they "provide us with two different snapshots of the world."[7] Two very nearly opposite, complementary worldviews.

Broadly speaking, the LH has a *physical worldview* that emphasizes matter, energy, logic, and details. The RH has an *intangible worldview* that favors images, insights, intuition, and context.

An Agatha Christie novel comes alive, for example, because of the LH's talent for understanding text and analyzing clues. This is complemented by the RH's gift for picturing characters and intuiting their feelings.

If you're stranded in the wild, the LH protects you by spotting subtle details—such as discerning poisonous berries from edible ones. The RH protects you by constantly surveilling the surroundings for any threats, as though it were armed with a pair of wide-angle binoculars.

As you're about to see, in this remarkable sibling cooperation and rivalry between the brain's fraternal twins, we can pinpoint the sources of enlightened and misguided faiths.

THE GREAT DIVIDE

During the 1960s and 1970s, neurosurgeons began severing the *corpora callosa* of their epileptic patients' brains in hopes of diminishing or even eliminating their seizures.[8] This daring technique worked well enough to earn its inventors the 1981 Nobel Prize in Physiology or Medicine.[9]

But the surgery also produced some shocking side effects in these split-brain patients. For example, a man named Joe.

When a photo of a frying pan is presented to Joe's right-hand field of view (apprehended by the LH), he readily sees and identifies it. But when it's shown to his left-hand field of view (apprehended by the RH), Joe sees nothing. *Absolutely nothing.*

But when he closes his eyes and draws whatever first comes to mind, he instantly sketches a frying pan—and doesn't know why![10]

Joe's strange experience underscores one of the most striking differences between the brain's fraternal twins. The LH *sees* the world, knows about it *consciously*, and can easily describe it *verbally*. The RH is *blind* to the world yet knows about it *unconsciously* and can describe it *nonverbally*.

This blockbuster discovery has enormous implications beyond just neuroscience. First, it demolishes the idea of "seeing is believing" and affirms the principle of "believing is seeing."

If you choose to believe only in what you can see and name, you're literally being half-brained. You're putting all your faith in what your brain's LH alone can see.

You're ignoring realities that are invisible to you, but that your brain's RH can somehow behold; realities like the frying pan photo Joe couldn't see and couldn't put into words, but that he was able to express *nonverbally*.

These are, I submit to you, precisely the translogical realities I described earlier in the book.[11]

Translogical realities are ineffable; you cannot see, prove, or even imagine them. But you can perceive them in their full glory by using your IQ *and* SQ. You can describe them by using your incomparable verbal and nonverbal human talents: art, music, and worship. And you can believe them with confidence because they are powered by *enlightened* faith.

Edward Young, the eighteenth-century English poet, once wrote: "By night, an atheist half-believes [in] a God."[12] Modern neuroscience gives new meaning to these nearly three-hundred-year-old words.

Given what we've learned about the brain, we now know for certain that even an Atheist's RH is unconsciously aware of things invisible to the eye—including, quite possibly, God. By day, through conscious effort, the Atheist can ignore his RH. But by night, asleep and disarmed, he'll have a harder time of it. Like Joe with his eyes closed, the Atheist is very likely to "see" past his wide-eyed blindness.

Second, from eavesdropping on the messages flying between the RH and LH via the *corpus callosum*, we've learned something astonishing: Most of the communication is not friendly. More often than not, one twin is barking at the other: "*Back off!* I've got this!"

"So much is this the case," explains Iain McGilchrist, renowned British psychiatrist and author of the excellent book *The Master and His Emissary: The Divided Brain and the Making of the Western World*, "that a number of neuroscientists have proposed that the whole point of the corpus callosum is to allow one hemisphere to inhibit the other."[13]

For some split-brain patients, this hidden civil war breaks out in surprising ways.

Listen to Vicki explain how her warring RH and LH made shopping at a supermarket a living nightmare. "I'd reach with my right [hand] for the thing I wanted, but the left would come in and they'd kind of fight," she says. "Almost like repelling magnets."[14]

Another split-brain patient was instructed to arrange blocks into a certain pattern. His left hand finished the job quickly and easily; but his right hand was confused and just couldn't do it.

When both hands were set to the task, they wrestled endlessly. "It was as if two people were fighting over performing this task," explains Michael Gazzaniga, the scientist in charge of the experiment and one of the world's leading split-brain neuroscientists. "One knew how, and one didn't, and one would fight for dominance of the situation."[15]

Here, then, in the supreme self-confidence (one might even say arrogance) of each fraternal twin . . . in its conviction that what and how it perceives reality is the most trustworthy . . . in the twins' spectacular rivalry and complementarity . . . *here* are the biological roots of *all* faith—IQ-based faith, SQ-based faith, misguided faith, and the most sagacious of all: enlightened faith.

As we've seen, the LH favors all things conscious, quantitative, and physical—the essential features of IQ-based faith. It's the faith that enables us to recognize and believe in truths and realities that can be seen, proven, and imagined. Truths and realities that are very important but, in the grand scheme of things, are trivial.

By contrast, we've seen that the RH favors all things unconscious, qualitative, and metaphysical—the essential aspects of SQ-based faith. It's the faith that enables you to recognize and believe in Gödelian truths and translogical realities that cannot be seen, proven, or imagined. It grants you access to, and gives you privileged glimpses of, life's most profound, deeply hidden secrets.

Misguided faith happens when you allow one fraternal twin to completely bigfoot the other. When you let it get away with saying: "*Scram, you idiot!* I've totally got this!" you lose the complementary value of both twins working together.

Specifically:

- **MISGUIDED IQ-BASED FAITH** is when you let your LH dominate or upstage your RH. When you ridicule the idea of God, for example, because it doesn't compute—it isn't logical.
- **MISGUIDED SQ-BASED FAITH** is when you let your RH dominate or upstage your LH. When you wish upon a star and expect the wish to come true because you believe in magic.

By contrast:

- **ENLIGHTENED FAITH** is when you insist that your warring brain halves get along. When you say to them, *"Play nice!"* and immediately discipline any half that threatens to bully the other—or that so much as gives it a dirty look.

Let me put it another way. Enlightened faith happens when you make full use of your IQ *and* SQ; when you put on those prized 3D glasses I spoke about earlier. When you keep them on, through thick and thin, till death do you part.

THE GREAT CATASTROPHE

I frequently run into people who believe that logic is somehow superior to faith; that logic is a virtue and faith a weakness. They believe that logical, computerlike reasoning is the most intelligent, most penetrating way to think.

These are folks who allow their LH to talk smack about their RH. Who allow their LH to screech at their RH, *"Scram, you idiot!"* without putting the bully in its place.

Tragically, this misguided, domineering-LH worldview is popular among well-educated men and women who should know better. Their half-blind worldview—their unchecked logical, material, positivist worldview—poisons conversations and cultures all over the industrialized world.

I call it *the great catastrophe.*

At the heart of this calamity is misguided IQ-powered reasoning that goes something like this:

Logic is superior to faith.
Science is based on logic and religion is based on faith.
Therefore, science is superior to religion . . . and the two will always be at odds.

One of the first full-throated defenses of these mistaken beliefs was penned by John William Draper, a nineteenth-century British-American polymath at the University of New York. In his landmark book *History of the Conflict Between Religion and Science*, Draper claims:

> The history of Science is not a mere record of isolated discoveries; it is a narrative of the conflict of two contending powers, the expansive force of the human intellect on one side, and the compression arising from traditionary faith and human interests on the other.[16]

Draper is dead wrong; science is *not* based on logic alone. As you will see in the very next chapter, science is guided by faith, just like religion.

Draper's thesis is also untrue because science and religion have coexisted peaceably, even synergistically, for most of human history. Their amicable relationship is well documented in countless published histories of civilization.[17]

Albert Einstein, though he revered science, didn't allow his LH to browbeat his RH. In a 1954 essay titled "Science and Religion," he writes:

> Science can only be created by those who are thoroughly imbued with the aspiration toward truth and understanding. This source of feeling, however, springs from the sphere of religion. . . . The situation may be expressed by an image: science without religion is lame, religion without science is blind.[18]

Einstein didn't believe in a personal God, which some Atheists play up to imply dishonestly that he was one of them. *He was not.*

In fact, Einstein greatly resented being called an Atheist. On one occasion, he complained that "there are yet people who say there is no God. But what really makes me angry is that they quote me for the support of such views."[19]

In an interview with the German-American poet George Sylvester Viereck, Einstein stated emphatically: "I am not an atheist."[20]

"Albert Einstein called the intuitive or metaphoric mind a sacred gift," notes author Bob Samples. For Einstein, "the rational mind was [merely] a faithful servant."[21]

Because of that, Samples adds poignantly, "It is paradoxical that in the context of modern life we have begun to worship the servant and defile the divine."[22]

In the movie *A Rumor of Angels*, Vanessa Redgrave's grieving character expresses a similar lament: "People invent machinery [e.g., powerful telescopes and microscopes] to improve on God's gift of perception, but refuse to credit the voices and visions that inspired them."[23]

Valuing logic over faith, as the great catastrophe does, is as misguided as valuing apple pies over apples. The former is a tasty invention; the latter, a natural, God-given gift.

Faith is a natural gift wired into your brain's twin hemispheres. It powers not only religion but also science—even my beloved disciplines of math, physics, and astronomy, as you're about to see.

Logic is a clever invention conceived by the Athenian philosopher Aristotle. It's a recipe for how to think in a certain, productive way—whose first and main ingredient is, as you're also about to see, *faith*.

Let me put it another way. Faith, the natural gift, does not depend on logic. But logic, the invention, depends on faith.

In reality, then, *faith is superior to logic*.

Why is it important for you to understand this? Why should you care what faith is, exactly?

You should care because of faith's titanic power. At its best, faith is not only mightier than logic; it's mightier than anything your brain can imagine.

Enlightened faith—the most potent kind of faith—is mightier than any other power in the universe.

Enlightened faith is more powerful than gravity, electromagnetism, the weak force, and the strong force because it's the foundation of science and technology, which has permanently empowered us to manhandle all of these forces.

Enlightened faith inspired our species to invent the rocket ship, which defeats gravity; dynamite, which works by commandeering the electromagnetic force; the X-ray machine, which works by subjugating the weak force; and the nuclear bomb, which works by marshaling the strong force.

Enlightened faith, in short, is the mighty genius that heeds both IQ's trivial rules of logic and SQ's profound, translogical voices and visions, and declares: "Believe!" And because countless people over the centuries have truly believed, our species—the amazing, unique *Homo sapiens sapiens*—can now literally move mountains.

If you have never thought about faith in this way—as the mightiest power in the universe—it's not too late. I urge you to start doing so right now.

This is not an abstract philosophical matter. By the end of this book, you'll see quite clearly that your joy, your sense of purpose, your very life, all depend on *what you believe*. Which itself depends on how well or poorly your brain's fraternal twins get along.

But if you're still skeptical—if you're someone who stubbornly believes that faith is a weakness and wants nothing to do with it—the next four chapters will explain how very mistaken you are. They deal with the scientific method, mathematics, physics, and astronomy—subjects in which I'm formally trained. Subjects widely misperceived as not needing faith, as being purely logical. Subjects that for some people are a religion—a tiny, misguided worldview called *scientism*.

7

HAVING FAITH IN THE SCIENTIFIC METHOD

God pity the man of science who believes in
nothing but what he can prove by scientific methods!
. . . For if ever a human being needed divine pity, he does.

JOSIAH GILBERT HOLLAND

I was born in the Mexican barrios of East Los Angeles to loving, hardworking parents. We had very little money, but I never felt poor.

Nearly all my relatives were laborers with at most a high school education. But I—*el bicho raro de la familia*, the family oddball—dreamed of becoming a scientist.

I was in love with the scientific method (SM). In my mind, it replaced blind faith, prejudice, and superstition with logic, objectivity, and truth. It was my savior, my best hope for escaping the barrio and having an exciting life.

My idols were scientists such as Ivan Pavlov, the Russian physiologist and Nobel laureate who believed that "the omnipotent scientific method . . . will deliver Man from his present gloom."[1]

And Karl Pearson, the British cofounder of mathematical statistics. He claimed there is "no way to gain a knowledge of the universe except through the gateway of scientific method."[2] Today, it sounds to me like Jesus saying, "I am the way and the truth and the life. No one comes to the Father except through me."[3]

Nowadays, the SM is more widely revered than ever. "For me, the scientific method is the ultimate elegant explanation," says Nathan Myhrvold, Microsoft's former chief technology officer. "It is the ultimate foundation for anything worthy of the name 'explanation.'"[4]

Alas for me, when I got to Cornell, I lost some of my naive veneration for the SM. For the first time in my young life, I saw how the sausage is actually made, discovered the shortcomings of both the SM and everyday practicing scientists.

By the time I headed to Harvard, I knew the SM for what it really is: a brilliant technique for understanding a tiny fraction of the physical world and virtually nothing about the world of metaphysics. It openly touts logic and objectivity but relies on faith in order to work. It's a method that's not even well defined.

In *Discourse on the Method of Rightly Conducting the Reason and Seeking for Truth in the Sciences*, René Descartes—the seventeenth-century French philosopher and co-formulator of the SM—describes the scientific method as consisting of four basic rules:[5]

1. **Be open-minded:** Approach science without any preconceptions or prejudices. Accept as true only those things that reliable experiments reveal to you.
2. **Be systematic:** Tackle the simplest mysteries first. Then systematically work your way up to the most vexing ones.
3. **Be analytical:** Break down every complex phenomenon to its simplest elements. Then, one by one, study each element.
4. **Be exhaustive:** When doing an experiment, take into consideration every relevant variable. Leave no stone unturned.

Descartes' four-step SM sounds quite reasonable. Yet search online for "scientific method," and you'll be served up scores of authoritative-looking websites explaining that the SM has five, six, seven, eight, or more steps—or even that it has no hard-and-fast steps at all.

Also, you'll find that the SM varies by discipline. Astronomy is done differently than biology is done differently than zoology, and so forth. Lab experiments, which happen in controlled environments, are done differently than field experiments, where variables are difficult to control.

Percy Bridgman, the Harvard-trained physicist and Nobel laureate, summed it up rather bluntly: "It seems to me that there is a good deal of ballyhoo about scientific method," he said. "Science is what scientists do, and there are as many scientific methods as there are individual scientists."[6]

Today—no longer that guileless, wide-eyed kid from East LA—I know what Bridgman means, and I agree with him. The SM is notoriously imprecise.

Sure, most scientists follow certain broad guidelines when designing experiments, collecting data, analyzing results, proffering explanations, writing them up, etc. But we do so with different gifts, styles, and idiosyncrasies, like musicians riffing in a jazz band.

It also doesn't help that the SM is not taught well, or at all. In my many years of formal scientific training, I was never once required to take a course on the scientific method. And it hasn't changed. Today, young scientists are still expected to *intuit* the SM and accept its validity with unquestioning faith.

In other words, science is a perfect illustration of "believing is seeing"—of faith preceding reason. You must believe in and diligently follow the SM in order to lay eyes on truths and realities not discoverable any other way. If you *don't* buy into the SM—imprecise as it is—you will not see those truths and realities. It's that simple. And that profound.

At its core, then, the SM is a *belief system*. It's not a wholly objective technique, contrary to claims made repeatedly by propagandists, such as the folks at the Geological Society of America (GSA).

In a publication titled *The Nature of Science and the Scientific Method*, the GSA makes these two fairy-tale claims: (1) "Science is non-dogmatic" and (2) "Science never requires ideas to be accepted on belief or faith alone."[7]

The GSA knows better, or ought to. Here's what I mean.

Dishonest claim #1: "Science is non-dogmatic." Science is, in fact, quite dogmatic. For a start, it insists that all card-carrying scientists accept the SM on faith, even without knowing exactly what it is. This dogmatism is not a bad thing, mind you. Science needs to police its adherents strictly, for exactly the same reason a religion does—to stave off an anarchy of beliefs and practices.

Science is also dogmatic about insisting that its members—and the public, too, actually—kowtow to its consensuses du jour. Skeptics, deniers, heretics—whatever you wish to call them—are not suffered gladly. Or kindly. You'll see what I mean in upcoming chapters.

Dishonest claim #2: "Science never requires ideas to be accepted on belief or faith alone." In fact, science indeed requires that we accept certain ideas *on faith alone*—starting with the claim that science is good for the world.

Think about it.

Science was invented to help explain the universe and our place in it and, therefore, to help create a better, healthier, longer existence for our species. Otherwise, what's the practical purpose of it all?

No one would deny that science has improved our lot in many significant ways. But it has also made our lives exceedingly more harrowing.

Thanks to science and technology, we now live under previously unheard-of threats—such as global environmental degradation, nuclear and biological weapons of mass destruction, and cyber warfare—any of which could swiftly lay waste to civilization.

These existential dangers have permanently destroyed our peace of mind. This is clearly evidenced by scores of studies documenting today's alarming rates of loneliness, depression, addiction, and suicide, especially among the world's young, rising generations.

States are now passing laws allowing kids to miss a certain number of school days strictly for stress or anxiety reasons—*mental health days*, they're being called. "In the age of social media, the internet, and constant bombardment of negative news and events," says Florida state representative Susan Valdes, "our children are bearing the brunt of a quickly-changing society."[8]

All told, the emotional and spiritual dysfunction caused or exacerbated by our scientific and technological progress (so-called) is staggering.[9] For now, therefore, we cannot possibly say the evidence supports the kind of blind faith science asks us to have in its goodness.

Science might ultimately lead us to a genuinely utopian world—I pray it does—but there is no way of *proving* it will. And there's plenty of evidence to suggest that science is actually helping to lead us astray—away from the kind of truth that really matters and toward a ghastly demise.

ARTICLES OF FAITH

Science is far more secularized today than it was in the days of its founding. Unlike then, today's SM adamantly forbids any explanation that references a deity or that smacks too much of the metaphysical. Only logical, material explanations are permitted.

But this does *not* mean that science sides with Atheism over other religions or that it says the material world is all there is. (Please pay close attention here.) *Atheists* claim that it does—and they want you to believe it—but it's not true.

What science says is that the material world is all that it wishes or feels qualified to explain. And it wants to offer strictly logical explanations of the material world's many mysteries. *Period.*

You can complain about science's decision to go secular, as many of my fellow Christians do. But like it or not, science has the perfect right to define itself—to enforce its *dogmas*—just as Christianity or any other religion does.

In any case, what hasn't changed is that science is still as faith-based as ever. Today's secularized SM still requires belief in axioms for which there is *evidence* but *no proof.*

To give you an idea of what I mean, here are three of science's axiomatic beliefs. They cannot be proved; it requires *faith* to go along with them.

1. The universe can be explained.

Science's belief in this axiom is rooted in the *principle of sufficient reason* (PSR), which claims that, given enough time, *everything* can be explained. Though mind-boggling, this belief appears to be true. "The eternal mystery of the world," Einstein marveled, "is its comprehensibility. The fact that it is comprehensible is a miracle."[10]

The PSR permits any kind of explanation; so, for many centuries, scientists published hypotheses salted with references to God without any concern. They saw science as the formal study of God's creation.

For them, the concept of God was entirely rational, even though he doesn't always behave in a trivial, logical way, as the Bible explains:

> "My thoughts are nothing like your thoughts," says the Lord.
> "And my ways are far beyond anything you could imagine.
> For just as the heavens are higher than the earth,
> so my ways are higher than your ways
> and my thoughts higher than your thoughts."[11]

For those early scientists, God was a profound, translogical reality—consistent with the best available evidence and therefore powered by enlightened IQ-and-SQ-based faith.

The PSR itself sprang from the mind of a devout Christian: seventeenth-century savant Gottfried Wilhelm Leibniz. He believed that "the recourse to an ultimate cause of the universe beyond this world, that is, to God, cannot be avoided."[12]

Al-Hasan Ibn al-Haytham, the eleventh-century Muslim polymath (also known as Alhazen), formulated an innovative seven-step SM "for gaining access to the effulgence and closeness to God." To understand God, he said, "there is no better way than that of searching for truth and knowledge."[13]

Ibn al-Haytham made it clear that his unprecedented experiments with light—done long before Isaac Newton's—were motivated by his open-minded, IQ-and-SQ-like passion to understand God's blinding glory. Between 1011 and 1021, he aggregated his spectacular results into his famous *Book of Optics*, a stunning, seven-volume treatise.

All in all, then, the main figures involved in designing the scientific method believed in the principle of sufficient reason—as well as the God of Abraham, Isaac, and Jacob. They believed his creation was a rational, explainable cosmos and that by studying it, they could better understand him.

Besides Leibniz and Ibn al-Haytham, these pioneers included Abu Ali Sina (Avicenna), Robert Grosseteste, Roger Bacon, Francis Bacon, René Descartes, Galileo Galilei, and Isaac Newton. In the fecund soil of these innovators' deep, spiritually intelligent Christian, Muslim, and Jewish convictions, science took root, sprouted, and grew into the mighty tree it is today.*

"*Metaphysics* is the root," avowed Descartes. "Physics the trunk, and all the other sciences the branches that grow out of this trunk."[14]

*Science sprouted up in many parts of the world, over many centuries, among many cultures and religions. But only in Christian Europe—beginning in the Middle Ages—did it take root and grow into today's mighty scientific method.

2. The simplest possible explanation is always the best.
Like the PSR, this dictum came from the mind of a Christian: the fourteenth-century English Franciscan friar William of Ockham. It's called Ockham's Razor.[15]

In his *Summa Logicae*, Ockham writes: "There is the argument that 'it is idle to bring about through several means what can be brought about through fewer.'"[16] In other words, why resort to a complicated explanation when a simpler one will do? Keep it simple, stupid!

Like all axioms, Ockham's Razor can't be proven; it must be accepted by faith. But since science has succeeded greatly by sticking with it, it's likely that believing in Ockham's Razor is a product of enlightened IQ-and-SQ-based faith.

Nevertheless, the axiom has some significant caveats.

First, it's not always easy to judge the simplicity of an explanation. For example, as we saw in chapter 3, our universe appears to be designed for life.

One possible explanation for the evidence is that there is, in fact, a Designer. Another possible explanation is that our universe is only one of countless universes. (Thus, it would not be a miracle if a lottery with lots of contestants produced at least one winner.)

Which explanation is simpler? Is the idea of a rational Designer more far-fetched than the idea of an infinite number of hidden universes?

I trust you see what I mean.

Second, if Ockham's Razor is true, *why* is it true? Why does nature demand simplicity? In truth, Ockham's Razor only broadens and deepens the mystery of the universe.

Third, Ockham's Razor does not apply to everyday life. My life rarely follows the simplest possible path; usually, it's just the opposite.

As the eminent religious scholar Huston Smith put it:

> The scientific method is nearly perfect for understanding the physical aspects of our life. . . . But it is a radically limited viewfinder in its inability to offer values, morals and meanings that are at the center of our lives."[17]

3. Doing experiments is the best way to understand the universe.

This is one of science's most cherished beliefs. Before the SM, natural philosophers relied on their IQ- and SQ-based intelligences alone. They'd sit in armchairs and argue among themselves about how the world works.

Scientists still do that—especially the theoreticians—but now they settle their arguments by doing experiments, all in accordance with the SM. It has made all the difference in the world.

As a boy, I used to read about Isaac Newton's pioneering prism experiments, Benjamin Franklin's kite experiments, and Louis Pasteur's polio experiments. The thrill of it—of concocting clever ways to pry secrets from nature—fueled my passion to become a scientist.

As a first-year grad student at Cornell, I finally got my wish. I was assigned to a major experiment intended to study subatomic particles, the pixels of matter.

It was exciting at first but, ultimately, disillusioning. I was disappointed to see that today's scientists are not like my childhood heroes: Experiments are no longer simple and straightforward.

The experiment to which I was assigned required a team of seventeen physicists, a warehouse full of complicated detectors, and a mammoth subterranean atom smasher located in Switzerland—which itself was operated by scores of other physicists and technicians from around the world.

What's more, the subjects of this gargantuan effort—subatomic particles—were too infinitesimal for me to ever see with my own eyes. We'd be having to infer their existence indirectly—*very indirectly*—from esoteric data.

It was a far cry from Ben Franklin capturing lightning in a bottle![18]

I quickly learned it wasn't just me and my humungous experiment that faced these issues. Scientists in every discipline are now far, far removed from the reality they claim to explain.

Paleontologists routinely draw extravagant, global conclusions about an entire species, based on the study of a single jawbone from a single individual dug up at a single location.

Astronomers make excited claims about the possibility of extraterrestrial life, based on exoplanets they cannot actually see but believe exist based on ever-so-subtle variations in the orbits and brightnesses of stars quadrillions of miles away.

Psychologists come to lavish conclusions about human nature and all people—young, old, rich, poor, rural, urban, educated, uneducated, black, brown, and white—based on studies primarily of white, college-age, paid volunteers.

And the troubling issues with today's empirical sciences don't stop there. We must also take into consideration the unavoidable uncertainties that come with designing, doing, and interpreting the results of modern scientific experiments.

DESIGNING AN EXPERIMENT

Designing an experiment begins with identifying some relatively simple part of a complex phenomenon—one that can realistically be measured.

Take Earth's climate, for example. It's a very, very complicated affair, with countless moving parts. To make sense of it, we must start by thinking *small and simple*—like measuring air temperature or sea levels or solar insolation (the amount of sunlight striking the ground) or cosmic radiation (yes, cosmic rays do affect the climate).

After deciding *what* to measure, we must figure out *how* to do it.

Measuring air temperature sounds easy enough, but it isn't. Do we use an old-fashioned mercury thermometer? A digital thermometer? An infrared thermometer?

And *where* do we make the measurements? Near asphalt, which heats up easily? Near water, which stays cool? On the ground? From space? You get the idea—there's nothing simple about designing even an experiment as seemingly straightforward as measuring temperature.

It's no wonder, then, that so many scientists get it wrong.

A team of investigators led by Malcolm Macleod, a neuroscientist at the University of Edinburgh, evaluated the designs of 2,671 experiments that involved testing promising new drugs on animals. These studies have life-and-death consequences for human patients worldwide.

Macleod's team found that the vast majority of these experiments failed—*failed*—in four key design areas: sample size, randomization, blinding (making sure neither scientist nor subject knows who's getting what), and conflicts of interest.

When the team focused on only those experiments done in the United Kingdom, the results were even worse. "It is sobering that of over 1,000 publications from leading UK institutions, over two-thirds did not report even one of [the] four items considered critical to reducing the risk of bias, and only one publication reported all four measures."[19]

DOING AN EXPERIMENT

Once we design an experiment—which can take months, even years—it must be submitted for approval from whoever is funding it. After that, it's time to roll up your sleeves and get to work!

I did my first real lab experiment when I was a freshman undergrad at UCLA. I won't go into the details, but it required radioactive strontium 90, an aluminum collimator, a Geiger counter, and endless hours of careful measurements.

I needed to have faith not only in my own abilities but in the equipment's reliability.

Back then, my equipment was simple enough that having faith in it was no big deal. Today in the experimental sciences, it's a very different ball game.

Today's experimentalists routinely place their faith not only in complex equipment they neither understand nor operate but in the legions of technicians who do. Everything from space telescopes and mass spectrometers to MRI machines and DNA sequencers.

Today's scientists also routinely place their faith in collaborators, whom they might know only superficially. And in still-wet-behind-the-ears grad students. And in the scores of government bureaucrats, industry bigwigs, university administrators, and eccentric, wealthy patrons who approve the experiments, pay the piper, and thus call the tune in many significant ways.

That's a whole lot of faith.

And there's shocking evidence that much of it is misplaced and misguided.

The prestigious British journal *Nature* conducted a survey of 1,576 scientists and discovered that "more than 70 percent of researchers have tried and failed to reproduce another scientist's experiments [one or more times], and more than half have failed to reproduce their own experiments."[20]

This colossal failure—now called the *reproducibility crisis*—afflicts research published in the world's most respected, peer-reviewed journals. Which means something is very wrong with how experimental science is being done.

DISCUSSING AN EXPERIMENT

After completing an experiment, it's time to analyze its results. This is a tricky business because the data—the so-called *evidence* or *facts*—can usually be understood in more than one way.

Inevitably, the published conclusion will be only one possible interpretation of the evidence—and maybe not even the best one. Perhaps the researcher has an unconscious bias; no one is perfect, after all. Like everyone else, scientists have preconceived notions, and conscious and unconscious biases, that inexorably influence their published conclusions.

In 1928, American anthropologist Margaret Mead published *Coming of Age in Samoa*, in which she presented her analysis of sexual customs in the South Pacific's Manu'a Archipelago. Her conclusions made Samoan society out to be laid-back, fun-loving, and peaceful—and its teenagers, sexually promiscuous and well-adjusted.

"In Samoa love between the sexes is a light and pleasant dance," Mead reported. "The idea of forceful rape or of any sexual act to which both participants do not give themselves freely is completely foreign to the Samoan mind."[21]

In fact, nothing could be further from the truth.

In 1983, Australian anthropologist Derek Freeman published a damning exposé titled *Margaret Mead and Samoa: The Making and Unmaking of an Anthropological Myth*. In it, he presents his own research on the Manu'a culture that completely discredits Mead's claims.

Manu'a natives, Freeman found, were a violent, jealous, and uptight people. During the time being reported, rape convictions were twice and twenty times greater than those in the United States and Great Britain, respectively.[22]

How did Mead get it so wrong?

Mead, an outspoken progressive, went to the islands already convinced that nurture trumps nature. She *believed* strongly that culture shapes people's behavior far more so than genetics—and that's what she *saw*. Her biased worldview corrupted her analysis.

Believing is seeing.

And that wasn't Mead's only failing.

It turns out she interviewed only young girls, no boys. And they lied to her about having free sex. Mead—an intelligent woman and a trained scientist—was taken in by the girls' whoppers because they confirmed her preexisting bias.

The ramifications of Mead's dishonesty and prejudice were enormous. As Freeman laments, her hugely popular, romanticized accounts of Manu'a life successfully "misinformed and misled the entire anthropological establishment."[23]

Though extreme, Mead's case is not an isolated one. A study by Italian scientist Daniele Fanelli discovered that an alarming *72 percent* of scientists knew of colleagues who had resorted to "questionable research practices." And 14 percent knew of colleagues who had outright falsified data.

Worse still, these results were based on self-reporting. So "it appears likely," Fanelli says, "that this is a conservative estimate of the true prevalence of scientific misconduct."[24]

THE FOUNDATION OF SCIENCE

Notwithstanding all these problems with experimental science, I still believe the SM is our most brilliant technique for understanding the physical world. I no longer see it as my savior—or humanity's savior—but I still have (guarded) faith in it.

Also, I'm not bothered by science's secularization. Science has every right to play by whatever rules it wishes. And as I explained earlier in the chapter, the SM's secularization does *not* mean that science denounces or has disproved the idea of God. Please remember that.

What does trouble me greatly are people—well-educated ones, even—who stubbornly refuse to acknowledge (or simply have never learned) that the SM was born and bred *by faith*, and that it continues to live and thrive *by faith*.

To those misguided souls, I say: *Wake up!*

As we've just seen, the axiomatic beliefs of science—the principle of sufficient reason, Ockham's Razor, empiricism, and countless others I don't have space here to name—are all *faith-based.* Science is a faith-based enterprise, not some trivial logical algorithm.

Without faith, science couldn't exist. And not just any faith—*enlightened* faith informed by both IQ and SQ, the result of the brain's LH and RH coming together in a powerful, irresistible way and urging us to *believe.*

Lots of other animals have IQ, but none has ever invented a scientific method because none has SQ.

SQ is the difference. Without SQ, you and I wouldn't have the slightest interest in anything beyond the Darwinian jungle; beyond surviving, procreating, and securing our next meal.

Without SQ, we wouldn't spend the time, money, and energy to design and deploy spacecraft to worlds quadrillions of miles away or dream about traveling to the stars.

Without SQ, we wouldn't read and write books such as this, musing about the meaning of life and swelling with pride for knowing the difference between a protein and protozoan, a quasar and a quantum, a legume and a light-year.

But you and I *do* have SQ. And because of that, our species alone invented science—undeniable, extraordinary evidence of our stunning uniqueness.

8

HAVING FAITH IN MATHEMATICS

Our minds are finite, and yet . . . we are surrounded by possibilities that are infinite, and the purpose of human life is to grasp as much as we can out of that infinitude.

ALFRED NORTH WHITEHEAD

If you desperately crave certainty in your crazy life, there's no better place to find it than on the fabled Isle of Mathematics. There, natives spend their days and nights producing the universally coveted antidote to uncertainty: 100 percent faith-free proof.

Right?

Wrong.

Some mathematicians no doubt dream of living on such an enchanted island. The late mathematician Julia Robinson liked to think of mathematicians "as forming a nation of our own without distinctions of geographical origins, race, creed, sex, age, or even time."[1]

But as a mathematician myself, I assure you there's no such thing as a faith-free proof. In fact, as you're about to see, the entire mythical Isle of Mathematics—*proof* itself—is supported by a vast sea of faith. Mathematics depends on beliefs that cannot be proven or imagined and, in many cases, are outright preternatural.

Mind you, it's risky to state this fact out loud—as a 2017 visitor to Stack Exchange, an online platform claiming to be "a question and answer site for those interested in the study of the fundamental nature of knowledge, reality, and existence," found out the hard way.

He asked the forum's math community this simple question: "Does mathematical proof require faith?" As an engineer, he hoped his query would stimulate a thoughtful discussion about a deep subject. "But," he said, "I was basically run out of town."[2]

Does mathematical proof require faith?

For an honest answer, journey back with me now to Athens and Alexandria twenty-four centuries ago. Meet a pair of young hotshots who, like Madonna and Prince, go by single names: Aristotle and Euclid.

Aristotle, poster boy for the brain's LH, has just invented the rules of logic, a strict new way of thinking. It's exemplified by something called a syllogism, which works like this:

All ravens are black.
Edgar is a raven.
Therefore, Edgar is black.

The first statement is an axiom, an assumption, a stated belief—hopefully an enlightened one. It cannot be proven, but if you have faith in it—if you *believe* it—you'll *see* something important about Edgar. Remember: *Believing is seeing*.

Awestruck by Aristotle's logic, young Euclid is at this very moment using it to deduce everything there is to know about *plane geometry*, the study of shapes on a flat surface.

To do this, Euclid must assume no fewer than thirty-three axioms—beliefs he cannot prove. Specifically, there are twenty-three definitions, five postulates, and five so-called common notions.

Here are the common notions:

1. Things that equal the same thing also equal one another.
2. If equals are added to equals, then the wholes are equal.
3. If equals are subtracted from equals, then the remainders are equal.
4. Things that coincide with one another equal one another.
5. The whole is greater than the part.

They're all pretty obvious, right? But ask yourself: *Why is that?* Why do these particular axioms seem so obvious to you?

It's because they're logical assertions you can easily verify for yourself by doing some simple counting experiments with marbles or beads. They're *trivial* truths that require only IQ-based faith to believe.

But some of Euclid's other axioms aren't so obvious, aren't so logical, aren't so trivial. In fact, they're downright stupefying and require SQ-based faith to believe.

For example, Euclid defines a *point* as something with no width, no depth, and no length. It's both something *and* nothing! Just like the quantum vacuum.

A point doesn't make sense. It's not logical. It isn't even something you can see or imagine.

I dare you to form a coherent mental picture of a point. You can't do it any more than you can picture something that is both alive and dead, black and white, true and false.

Yet Euclid's definition of a point isn't nonsense; you can't just dismiss it. Why? Because it's a profound, *translogical*, SQ-based axiom, which the young innovator is using to prove the entirety of plane geometry.

Is your head spinning?

Think back to Joe, the split-brain patient we met in chapter 6. He couldn't see the photo of a frying pan put in front of him. But when he closed his eyes, his brain's RH—powered by its uncanny intuition, its SQ—spoke to him wordlessly; and instantly Joe sketched a frying pan, albeit a crude approximation of one.

The very same phenomenon is in play here. A Euclidean point is something your eyes cannot see, your IQ cannot comprehend, and your imagination cannot even fathom. But your brain's RH—powered by SQ—is able to perceive it sightlessly and wordlessly; and with its help, and that of a sharpened pencil, you can draw a crude approximation of it.

Return with me now to the present.

Euclid's SQ—his religious-like faith in axioms that aren't logical and can't be seen or even imagined—proved to be so enlightened that he revolutionized mathematics.

For more than two thousand years, we've used Euclid's historic achievement—the countless theorems of his plane geometry—to successfully build bridges and skyscrapers, lay out floor plans, and compute trajectories to the moon and worlds beyond. For that reason, more geometry textbooks have reportedly been sold and read worldwide than any other book except for the Bible.

FAITH IN LOGIC

Euclidean geometry is stunning evidence that logic founded on enlightened faith can be powerful. But are there limits to the power of logic? Or is it all-powerful, as many today would have you believe?

The answer is no, logic is *not* all-powerful. It has limits. Severe ones, it turns out.

The head-spinning saga of logic's epic takedown began with the brilliant German logician Friedrich Ludwig Gottlob Frege. He was a Euclid wannabe. He yearned to do for arithmetic (the study of numbers) what Euclid had done for geometry (the study of shapes).

Frege began by articulating six axioms—far fewer than Euclid's thirty-three. These were beliefs he was sure would let him prove all the theorems in arithmetic, starting with 1+1 = 2. Frege toiled away for years before, finally, in 1893, publishing the first installment of what he believed would be a three-volume magnum opus: *The Basic Laws of Arithmetic.*

Nine years later, Frege finished the second volume. But just as he was about to submit the manuscript to the publisher, he received some bad news from the legendary Welsh mathematician Bertrand Russell.

In a letter dated June 16, 1902, Russell informed Frege that he'd found an error in the first volume. Not a typo but a devastating flaw in Frege's logic.

I won't get into the technicalities, but here's the gist of it: Russell found a problem with Frege's reasoning that traced back to his fifth axiom, which defines the membership of a group, or set. Sets are a big deal in math and must be carefully defined.

For instance, the group of all living persons on Earth is the set that contains *all* and *only* those people on Earth who are not dead. Simple enough, right?

But now imagine a village of clean-shaven men. Among them is the village's one and only barber, who boasts: "I shave *all* and *only* men who do not shave themselves."

Question: What can you say about *the set of all men who shave themselves*? Does the set include the barber? In other words, is the barber a man who shaves himself?

Don't answer too quickly; it's tricky.

Suppose you say, *yes*, the barber shaves himself. That would contradict the barber's boast that he shaves *only* men who do not shave themselves.

Suppose you say, *no*, the barber does not shave himself. That would contradict the barber's boast that he shaves *all* men who do not shave themselves.

There is, then, no logical answer to my question. Logic utterly fails to crack the mystery.

Here's another way of seeing the problem. Question: Is this headline true or false?

THIS STATEMENT IS FALSE.

If the sentence is true, it's false. If it's false, it's true!

Once again, logic spins out of control, sucking us into a maelstrom of circular reasoning from which there is no escape.

That, in essence, is the flaw in Frege's logic that Russell detected. As you can imagine, the Euclid wannabe was flabbergasted.

"Your discovery of the contradiction has surprised me beyond words," Frege wrote back to Russell. "And, I should almost like to say, left me thunderstruck, because it has rocked the ground on which I meant to build arithmetic."[3]

In 1903, Frege went ahead and published his second volume anyway but with this mournful disclaimer: "Hardly anything more unfortunate can befall a scientific writer than to have one of the foundations of his edifice shaken after the work is finished. This was the position I was placed in by a letter of Mr. Bertrand Russell, just when the printing of this volume was nearing its completion."[4]

Frege never published the third volume of his failed masterpiece, nor much of anything else for sixteen long years. By 1923, he had totally given up on the idea of using logic to prove arithmetic, and he died on July 26, 1925, in relative obscurity—although today he's rightfully lauded for being an extraordinary logician.[5]

All mathematicians, not just Frege, were unnerved by this horrific turn of events. By 1925, David Hilbert, arguably the century's most gifted mathematician, conceded that "the present state of affairs where we run up against the paradoxes is intolerable. Just think, the definitions and deductive methods which everyone learns, teaches, and uses in mathematics, the paragon of truth and certitude, lead to absurdities! If mathematical thinking is defective, where are we to find truth and certitude?"[6]

Where, indeed?

Hilbert urged his colleagues not to lose hope. The paradoxes, he said, were merely a symptom of faulty axioms—Frege's fifth axiom being one such culprit. The solution, therefore, was to select axioms more carefully . . . ones that were enlightened, not misguided . . . ones that were, in a word, *self-consistent.*

Mathematicians responded enthusiastically to Hilbert's plaintive battle cry, including Bertrand Russell and his brilliant senior colleague Alfred North Whitehead. In 1910, they published (shades of Frege!) the first volume of a planned three-volume work, titled *Principia Mathematica.* It promised to pave the way at last for putting mathematics back onto a solid, logical footing—free of any paradoxes.

Principia was hailed by mathematicians far and wide—until it came under the scrutiny of Kurt Gödel. In 1931, the reclusive genius spotted a major glitch—not just in *Principia* but in logic itself. A fundamental deficiency that Gödel proved could not be overcome. Not ever.[7]

Gödel's bombshell was published in two parts, now commonly called the incompleteness theorems. Here's a slightly technical way of describing them:

> Given any formal system of logic powerful enough to describe all the truths in arithmetic, it will be either *incomplete* (there will be truths it cannot prove) or *inconsistent* (it will be infected with paradoxes and therefore totally unreliable).

Is your mind spinning again? It should be.

Here's a simpler way of explaining Gödel's theorems. Whenever you try thinking logically about a complicated subject, one of two things will *always* happen:

> **Possibility #1:** You'll say and believe something that's genuinely true, but you'll never be able *prove* it. No matter how hard you try, logic will fail you because logic is not powerful enough to do the job.

The late Swiss-American logician Verena Huber-Dyson put it this way: "There is more to *truth* than can be caught by *proof*."[8] I prefer simply saying, *Truth is bigger than proof.*

> **Possibility #2:** You'll *prove* something is true using seemingly airtight logic; but in fact, it's not so. Even though your logic seems rigorous, it isn't; it's riddled with stealthy paradoxes.

As Morris Kline, the renowned American mathematician, wrote in his textbook *Mathematics for the Nonmathematician*, "Logic is the art of going wrong with confidence."[9]

Possibility #2 was Frege's problem—and now, as Gödel proved, it was Russell's and Whitehead's as well. How bitterly ironic.

A woeful Russell quickly recognized that he had unintentionally both enabled and validated Gödel's devastating achievement. Three decades earlier, Russell had found a loose thread in Gottlob Frege's work; now Gödel had yanked on it good and hard. In doing so, he'd unraveled not only the *Principia* but the very concept of mathematical proof.

Russell was floored—and not just as a mathematician. He'd always been a devout, outspoken Atheist, gleefully bashing Christianity and other religions—much as Richard Dawkins does today. In 1927, he'd even gone to the trouble of penning an essay titled *Why I am Not a Christian*, wherein he offers an eloquent defense of his Atheism.

For that reason, you could say Russell was a walking, talking logical paradox. While savaging religions other than Atheism, he had placed fanatical, religious-like faith in logic and mathematics. A misguided faith, it turns out, that Gödel had now thoroughly and irreparably demolished.

"I wanted certainty in the kind of way in which people want religious faith," an elderly Russell mourned in *Portraits from Memory*. "I thought that certainty is more likely to be found in mathematics than elsewhere. But . . . after some twenty years of very arduous toil, I came to the conclusion that there was nothing more that *I* could do in the way of making mathematical knowledge indubitable."[10]

Today, the implications of Gödel's incompleteness theorems ripple well beyond the shores of the Isle of Mathematics. Here are three illustrations.

First, Gödel's theorems seriously undermine the belief in a theory of everything (TOE)—for example, the grand unified theory (GUT), the Holy Grail of physics. GUT aspires to provide a single, coherent explanation of the four known forces in nature: *gravity*, *electromagnetism*, the *strong force*, and the *weak force*.

Einstein spent his final years doggedly seeking out a GUT, but he failed. Indeed, by proving that logic is not powerful enough to describe arithmetic, let alone the universe, Gödel's theorems teach us that pursuing any kind of *logically self-consistent* TOE is as delusional as believing in the Tooth Fairy.

Second, Gödel's theorems easily allow for the statement "God exists" to be true but unprovable logically. Remember: *Truth is bigger than proof.*

Put another way: If logic can't even tackle arithmetic without losing its mind, it stands no chance of ever settling an argument about God, a subject just a tad more complicated than 1 + 1 = 2.

Third, Gödel's theorems affirm that the Isle of Mathematics floats in a sea of faith. In order to do business, mathematicians must first and foremost *believe* in unproven and possibly unprovable axioms.

No mathematician, however brilliant—not Euclid, Frege, or Russell—can create a logical argument without first throwing his or her *faith* behind a set of assumptions.

At worst, the assumptions will be powered by misguided faith and lead to disastrous results—as Frege and Russell painfully discovered. At best, they will be powered by enlightened faith—as Euclid and Gödel found out.

Either way, Gödel's theorems *prove*—using logic—that mathematics is a discipline founded on faith. In principle, no different from any religion.

"If a 'religion' is defined to be a system of ideas that contains unprovable statements," observes John Barrow, the eminent Cambridge University mathematician, in *The Artful Universe*, "then Gödel taught us that mathematics is not only a religion, it is the only religion that can prove itself to be one."[11]

FAITH IN AXIOMS

Even though mathematicians admired Aristotle's logic and Euclid's geometry, they wondered: Is it the *only* logic possible? Is it the *only* geometry possible?

The answer is *no*—there are many possible logics and many possible geometries. Each one depends on a different set of axioms, or beliefs.

One of Aristotle's core beliefs is called the *principle of the excluded middle* (POEM). It believes that something is either true or false; there is no in-between.

However, it turns out there are many rational alternatives to the

POEM, and mathematicians have used them to produce a veritable smorgasbord of non-Aristotelian logics. For instance:

1. **Three-valued logic** is based on *believing* that something can be true, false, or unknown.
2. **Four-valued logic** is based on *believing* that something can be true, false, true *and* false, or unknown.
3. **Fuzzy logic** (yes, that's what it's called) is based on *believing* that something can have an infinite number of truth values. That is, something can be anywhere from zero to 100 percent true.

   Fuzzy logic is used to program electronic devices that must react to sudden developments in nuanced, non-black-and-white ways—for example, the computer chips that control antilock brakes.

   The fuzzy-minded chips must decide how hard to hit the brakes by weighing the truth values of many different factors. They include "the car's speed, the brake pressure, the brake temperature, the interval between applications of the brakes and the angle of the car's lateral motion to its forward motion."[12]

The discovery of so many different kinds of logic has long since dethroned Aristotle's magnificent achievement. "The logic of Aristotle," the late Princeton University mathematician Edward Nelson explained, "is inadequate for mathematics. It was already inadequate for the mathematics of his day."[13]

Ouch.

There's a lot of talk in public these days about the importance of critical thinking—of teaching students to be critical thinkers. I agree with this wholeheartedly.

But given the astonishing developments in mathematics, you

need to understand that *critical* thinking now means something far more diverse and complex than merely *logical* thinking.

Aristotle's original recipe is just one of countless ways to reason wisely, to get at the truth reliably—and it's not even the strongest or most revealing. Among those in the know, ordinary Aristotelian logic comes under the category of a two-valued crisp logic—the Model T of critical thinking.

A similar fate has befallen Euclid's geometry.

One of Euclid's core axioms is that parallel lines never cross, even when extended to infinity.

However, this is true only for flat surfaces, not for curved surfaces. That single realization has produced an explosion of non-Euclidean geometries. For instance:

- **Spherical geometries** apply to rounded surfaces, like Earth. Lines of longitude play the role of parallel lines, which gradually bend toward one another and converge at the poles.

- **Hyperbolic geometries** apply to saddle-shaped surfaces. On these worlds, parallel lines diverge—like a flock of birds peeling away from each other.

- **Riemannian geometries** (named after their nineteenth-century German inventor, Georg Friedrich Bernhard Riemann) apply to surfaces with four, five, six, and more dimensions. These surfaces can be flat, spherical, or hyperbolic.

Multidimensional Riemannian surfaces are impossible for us to see, or even imagine—just as with Euclid's profound, translogical concept of a point. They're spectacular, unfathomable products of human SQ.

Moreover, like Euclid's *point*, Riemannian surfaces have proven to be quite useful. Einstein uses a 4D Riemannian surface in general relativity to describe the behavior of gravity. It has three space dimensions (up/down, right/left, forward/backward) and one time dimension.

Over the years, experiments have repeatedly affirmed Einstein's theory. This implies that not only the theory but also the Riemannian surfaces it relies on are the creation of enlightened IQ-and-SQ-based faith.

The stunning success throughout history of scientific theories that rely on far-out, translogical mathematical axioms and concepts—be it a 4D Riemannian world, geometrical point, quantum vacuum, virtual particle, and so forth—astonished Einstein. "How can it be that mathematics, being after all a product of human thought which is independent of experience, is so admirably appropriate to the objects of reality?"[14]

Years ago, while attending a physics seminar at Louisiana State University, I spent a few days with Eugene Wigner, the legendary Hungarian-American mathematician and Nobel Laureate.[15]

Wigner knew nothing about my speculations concerning SQ; but in his own way, he recognized that mathematics appears to be powered not just by the rules of logic, not just by machine-like ways of thinking, but by the revelations and whisperings of a superintelligence that transcends mere IQ.

In an essay published in 1960, Wigner observed that "the enormous usefulness of mathematics in the natural sciences is something bordering on the mysterious and . . . there is no rational explanation for it."[16]

FAITH IN PROOF

There was a time when mathematical proofs were brief enough to be checked and double-checked by hand. The hundreds of proofs I did in high school geometry class were like that—and my teacher docked my grade for every step of a proof I got wrong. I didn't realize it back then, but being exposed to such fanatical rigor was the true start of my scientific training.

Proofs in high school geometry class are still like that: short and sweet. But in professional, high-level mathematics, those days are long gone.

There isn't a single inflection point in time you can point to and say, "*There!*—that's when the calamitous change happened." But by 1993, when *Scientific American* ran a story by veteran science journalist John Horgan titled "The Death of Proof," it was crystal clear the days of crystal clarity in mathematics had gone the way of the dinosaurs.

"For millennia, mathematicians have measured progress in terms of what they can demonstrate through proofs—that is, a series of logical steps leading from a set of axioms to an irrefutable conclusion," wrote Horgan, sounding the death knell. "Now the doubts riddling modern human thought have finally infected mathematics."[17]

Horgan's article was published in 1993 because that's when British mathematician Andrew Wiles claimed to have proven a 350-year-old mathematical mystery known as Fermat's Last Theorem. The alleged proof was hundreds of pages long, so it was no easy task to vet it. Worse, when mathematicians finally succeeded in doing so, they spotted a serious error.

For Wiles, it was back to the drawing board.

A year later, when he claimed to have corrected the error, his colleagues naturally were skeptical. But after going over the final, lengthy proof with a fine-tooth comb, a jury of respected mathematicians affirmed Wiles's historic effort, and it was published in 1995.

I did a story about it for *Good Morning America*. I flew to Princeton University and interviewed the man whose proof heralded a new age in mathematics. An age afflicted with growing uncertainty.

Since then, the crisis has only worsened. Proofs aren't just longer and harder to verify than ever; more and more of them are being generated with the help of computers. This is introducing into modern mathematics even more uncertainty and unverifiability.

"I think that we're now inescapably in an age where the large statements of mathematics are so complex that we may never know for sure whether they're true or false," says Keith Devlin, a British mathematician at Stanford University. "That puts us in the same boat as all the other scientists."[18]

As of this writing, the longest mathematical proof in history was produced in 2016 by three humans and a supercomputer that resides in an 11,000-square-foot building at the University of Texas at Austin. The supercomputer, Stampede, is an electronic behemoth that feeds on three megawatts of electric power.[19]

Stampede, et al.'s record-setting proof is *200 terabytes long*. That's the digital equivalent of the entire US Library of Congress. It would take ten billion years (nearly the age of the universe) simply to read it—and even longer to validate each step.[20]

The Sea of Faith upon which the mythical Isle of Mathematics now floats, then, is far deeper and broader than anyone could have imagined. Certainly, neither Aristotle nor Euclid foresaw any of what has happened.

But that's bad news only for people who were misguided enough to believe in the existence of 100 percent faith-free proof. No such thing exists or ever has existed on the Isle of Mathematics, just as Bigfoot does not and never has existed anywhere.

That's right. As we've just seen, the Sea of Faith on which mathematics floats teems with unprovable, unimaginable, translogical, SQ-based beliefs.

Beliefs that defy mere IQ.

Beliefs that describe the real world with uncanny truthfulness.

Beliefs you must embrace if you wish to behold their astonishing revelations.

9

HAVING FAITH IN PHYSICS

Not only is the Universe stranger than we think,
it is stranger than we can think.

WERNER HEISENBERG

Have you ever seen a large bird take off? It trots along the ground faster and faster, flapping its mighty wings, placing one foot in front of the other. Then comes that magical moment when it lifts off.

That's us when we combine IQ with SQ.

With our IQ, with logic, we trot along at a good clip—one step at a time—toward the solution of a problem. But with our SQ, with our natural cognitive superpower, we flap our mighty wings and take flight. We soar to dizzying summits inaccessible by foot.

Our SQ takes us to faraway places that our IQ can't see, can't prove exist, can't even imagine. It's the engine of humanity's unparalleled curiosity.

It reminds me of a song lyric popular right after World War I, when American soldiers—many of them farm kids—were returning from exotic locations overseas. "How ya gonna keep 'em down on the farm," it went, "after they've seen Paree?"[1]

SQ's unfathomable, penetrating intelligence opens your eyes to exotic, translogical realities. Once you use it, nothin' is gonna keep you down on the ol' Darwinian farm ever again.

Because of our SQ, we alone are curious about gravitational waves, neutrinos, and the quantum vacuum—things having no obvious bearing on our gritty struggle to survive. In fact, our hot pursuit of such esoteric phenomena has produced dreadful inventions that seriously *imperil* our existence.

Why are we wired to be curious about the remotest, most arcane mysteries of this world? And about what might lie *beyond* this world?

It's not a flaw in our design, as some Atheists claim. It's clear evidence that such exotic realities do indeed exist—realities that can't be seen, proven, or even imagined.

No one knows that better than my fellow physicists.

PHYSICS GROWS UP

Physics, science generally, is like a perpetual teenager: It never stops growing. That's both good news and bad news.

It's good because physicists are constantly learning new and surprising lessons. It's bad because it means scientific theories are perishable. New discoveries continually upend the status quo, ousting whatever misguided thinking and misguided faith might be tainting the physicist's worldview at the time.

Occasionally, the new lessons force major makeovers of the physicists' worldview—jarring growth spurts, if you will. Here are the four biggest spurts it's experienced so far:

Growth Spurt #1: In the fourth century BC, the Greek philosopher Aristotle produced a series of manuscripts collectively titled *Physics* (Φυσικὴ ἀκρόασις). It was the world's first physics textbook, the beginning of Aristotelian physics.

Growth Spurt #2: In 1687, the renowned Englishman Isaac Newton published *Mathematical Principles of Natural Philosophy* (*Philosophiæ Naturalis Principia Mathematica*)—commonly called Newton's *Principia*. It was the beginning of Newtonian or classical physics.

Growth Spurt #3: At the start of the twentieth century, Albert Einstein and a legion of others published ideas about the innermost structure of matter and light. It was the beginning of quantum physics, a major branch of modern physics.

Growth Spurt #4: Also at the start of the twentieth century, Einstein produced the theories of special and general relativity. This was the beginning of relativistic physics, another major branch of modern physics.

If I were to liken physics to an animal, it would not be a snake, which sheds its old skins as it grows. It would be a nautilus, which continually and elegantly expands the size of its shell as it grows.

As physics grows, it doesn't completely jettison its old beliefs; it amends, reinterprets, and adds to them. Thus, modern physics retains vestiges of even Aristotelian physics.

Case in point: Aristotle allowed gods into his explanations of the natural world. In *Metaphysics*, he writes:

> Our forefathers in the most remote ages have handed down to their posterity a tradition, in the form of a myth, that these bodies [sun, moon, planets] are gods, and that the divine encloses the whole of nature. . . . One must regard this as an inspired utterance.[2]

Modern physics no longer allows such overt references to gods in its explanations, but in recent decades especially, it has increasingly resorted to concepts that smack of the metaphysical. In fact, as I'm about to explain, many of these concepts are more fantastical than anything you'll find in Aristotle's physics—or even Lewis Carroll's Wonderland.

PHYSICS IN WONDERLAND I

On June 9, 1905, the prestigious German scientific journal *Annalen der Physik* (*Annals of Physics*) published a paper submitted by a young, unknown physicist named Albert Einstein. The title sounded mundane and esoteric: "On a Heuristic Point of View about the Creation and Conversion of Light."[3]

It turned out to be the equivalent of the American Revolution's famous shot heard round the world. The moment when physics underwent its biggest transformation since Aristotle and Newton. The moment when physics stepped through the Looking-Glass.

We call it quantum physics. "It was so revolutionary," remarked the legendary Isaac Asimov, "that almost no physicist . . . could bring himself to accept it."[4]

I don't have enough space here to fully explain quantum physics. And even if I did, I still couldn't do it.

No one can fully explain quantum physics. Not even the late Nobel laureate Richard Feynman, who in *The Character of Physical Law* candidly confessed, "I think I can safely say that nobody understands quantum mechanics."[5]

Quantum physics (aka quantum mechanics) is a work in progress. A half-baked cake. An exceedingly strange child still growing up.

Quantum physics offers us teasing glimpses—but not a complete view—of Wonderland's microscopic weirdness. It offers explicit equations but conflicting interpretations of the equations.

It's like a coin that falls between the couch cushions. The more you reach for it, the more it eludes your grasp.

That's quantum physics in a nutshell.

To see more clearly what I mean, take a look at five shocking claims that quantum physics makes about Wonderland—claims well beyond our ability to fully comprehend. They defy words, they defy proof, they defy even the human imagination. Yet physicists *believe in them* the way Aristotle once believed in gods.

1. All residents of Wonderland are paradoxical.

I already explained in chapter 4 that rays of light are paradoxical. They behave simultaneously like particles *and* waves, which is akin to claiming that something is simultaneously black *and* white.

No one can fully understand such a flagrant logical contradiction, not even Einstein himself.

In 1954, a year before he died, Einstein confessed to his old friend Michele Besso, "The whole fifty years of conscious brooding have not brought me nearer to the answer to the question, 'What are light quanta?' Nowadays every scalawag believes he knows what they are, but he deceives himself."[6]

And the translogical strangeness doesn't end there.

An electron, for example. Physicists used to believe it was a particle, plain and simple. But now we know that it, too, behaves simultaneously like a particle *and* a wave. The discovery of that profound, translogical reality is credited to a French grad student of noble birth named Louis-Victor Pierre Raymond de Broglie.[7]

These translogical entities don't just inhabit a far-removed, Wonderland-like quantum realm. They're the stuff you and I are made of!

At your very core, in other words, you are a living, breathing logical contradiction. You are a translogical, metaphysical creature, more strange, more inscrutable, more unimaginable than your own mind can fully fathom.

2. Some residents of Wonderland can teleport and communicate instantaneously.

To go from the ground level of a tall building to its twentieth floor, you need to take an elevator or climb the stairs. Even if you were Superman or Superwoman, you'd still need to cover the distance between floors.

Right?

In Wonderland, not necessarily. For instance, an **atom** is like a skyscraper with many floors. According to quantum physics, electrons can—and do—go from one floor to another *without* traversing the floors in between. Quite literally, they dematerialize from one floor and instantaneously rematerialize on another.

Then there's the business of communication. Normally, it takes time to deliver a message from A to B. Even today's so-called instant messaging takes time to get from transmitter to receiver.

But not necessarily in Wonderland. For instance, an **electron** spins like a top, either clockwise or counterclockwise. Suppose two electrons from the same atom fly apart like twins separated at birth, each with a certain spin. Measuring the spin of one electron instantaneously affects the spin of the other—as if the two electrons are in telepathic communication. We call this mysterious behavior *quantum entanglement.*

Quantum entanglement has been corroborated by many experiments. In one recent study, scientists at the University of Science and Technology of China in Hefei found that twin light quanta separated by a record 746 miles still seemed able to communicate with each other instantaneously.[8]

3. Residents of Wonderland can exist in many places at once.

When I was in middle school, my science teacher told us that an atom is like a miniature solar system. At its center is a nucleus made of neutrons and protons; whirling around the nucleus, like so many planets, are electrons.

Those days are long gone.

According to quantum physics, an atom is all *smeared* out, like a giant wave. But not like a wave at the beach. It's like a *wave of probability*.

The odds of an atom being at any one place are quite high. *But* . . . there's a chance, albeit infinitely small, that it's somewhere else—maybe even across the universe.

At any given moment, a typical atom is probably where classical physics says it should be. But you can't say it's there with complete confidence because it can be in many places simultaneously.

That's true for *any* typical microscopic resident of Wonderland, not just atoms. It has the odd property of being smeared out in more than one place at once.

Consider, for example, an electron confined inside a well—like a marble trapped in a ditch. You'd think it couldn't ever get out on its own, but it can. Why? Because while most of it is indeed confined inside the well, some part of it exists within *and* outside the walls of the well.

I know it sounds crazy—Wonderland is a crazy place—but because of the electron's smeared-out existence, there's a small chance that, if you wait long enough, it will materialize outside the well, as if it were a spirit passing through a solid wall. We call this astonishing phenomenon *quantum tunneling*.

4. Experiments in Wonderland are never truly objective.

Generally speaking, physicists do their level best not to allow biases and sloppiness to despoil the results of their experiments. But according to quantum physics, objectivity is something of a myth.

The reason: Physicists and their equipment cannot avoid interacting with whatever they're observing. That interaction inevitably affects the observation.

This unintended corruption is especially significant when

observing tiny things, such as light quanta and subatomic particles. But it ultimately afflicts *all* experiments.

Let me say it differently.

You've heard the old saying, "Beauty is in the eye of the beholder." It means that *beauty* has no meaning independent of the observer. Beauty and its observer are inextricably connected.

Likewise, it can be said that the entirety of Wonderland has no meaning without observers to describe it. Wonderland and its observers, its sentient inhabitants—that's us!—are inextricably connected.

"The Universe could only come into existence if someone observed it," says Sir Martin Rees. "The Universe exists because we are aware that it exists."[9]

Objectivity, then, is a myth.

You and I are inevitably part of our descriptions of Wonderland.

There is no avoiding this quantum corruption.

Here is a concrete example of what this means: Imagine a spinning electron locked inside an opaque box. According to classical physics, the electron is either spinning clockwise or counterclockwise; there is no third possibility. Moreover, if you leave the electron alone, the direction of its spin will not change.

Quantum physics sees things very differently. According to quantum physics, an electron behaves like *a wave of probability*. It is simultaneously spinning clockwise *and* counterclockwise.

However . . . the instant you open the box, the instant you interact with the electron, you *cause* one of the possibilities to be fully realized. The electron you observe will be spinning in only one direction—either clockwise or counterclockwise—and no longer both.

Thus, you are *not* an objective observer.

You are *not* merely observing some preordained reality.

You are unavoidably helping to determine the outcome.

You.

It's an *absolute truth*, then, that no matter how carefully or honestly you do an experiment, its results will always be tainted by your involvement in it.

"Science no longer is in the position of observer of nature," explained the German physicist Werner Heisenberg, "but rather recognizes itself as part of the interplay between man and nature."[10]

Here's another way of wrapping your mind around this: The behavior of a woman you've never met is represented by a cloud of many possibilities. But the instant you meet her, she becomes *one* of those possibilities.

Also—and this is the essential point—*your* interaction with her influences the outcome of your meeting. You affect how she behaves.

Someone else could very well experience a different outcome from meeting the same woman. Someone else could elicit from her completely different behavior.

So . . . going back to that electron. When I open the box, I may or may not experience the same outcome that you would. There's a fifty-fifty chance I will. And a fifty-fifty chance I'll get the opposite result.

Same electron. Same box. Opposite result.

If enough people open the box, the outcomes will average out to fifty-fifty. Half of the observers will see the electron spinning clockwise; half, counterclockwise.

We call this intriguing phenomenon the *quantum measurement problem*. Why problem? Because we don't really understand how reality can be so subjective.

It's easy to understand how the world of opinion is subjective. But hard-core *physical reality*?

"How do you go from the fuzzy, hazy reality of quantum probabilities, where things can be in many locations with different likelihoods, to the single, definite reality we observe when we do a measurement?" asks Columbia University physicist Brian Greene. Answer? "We don't know."[11]

It's a problem all right.

5. We can never know everything about Wonderland. Never. Think about all the measurements you make in life. You step on a scale to weigh yourself. You size up your living room windows for new curtains. You carefully dole out the ingredients of a recipe. You track the height of your growing child.

For physicists, measurements are *everything*.

Earth's temperature. A glacier's mass. The diameter of a satellite's orbit. The age of the universe. Physicists must get their measurements right every time if they hope to get published, be believed by the public, earn funding, and win prestigious awards.

Back in the days of Aristotle and Newton, physicists believed there was no limit to how accurate a measurement could be. They had faith that one day—with the right combination of cleverness, skill, and equipment—they could nail down every little thing about the universe with unlimited precision.

Then 1927 happened.

In that fateful year, Werner Heisenberg published a mathematical expression called *the uncertainty principle*. It claims that no matter how hard we try, no matter how clever or skillful we are, we can never find out everything about the universe.

It's impossible to speak about Heisenberg's uncertainty principle without recalling Gödel's incompleteness theorem. Both were shocking discoveries of insurmountable limits (of the scientific method and logic, respectively).

In L. Frank Baum's *The Wonderful Wizard of Oz*, a tiny wizard controls the land of Oz from behind a curtain. This is essentially what the uncertainty principle says about our own Wonderland.

It says that the innermost secrets of the universe lie hidden behind a curtain of uncertainty. A curtain that physics, unlike Dorothy, can never pull away—and that even our super-powerful SQ can peek through with only limited success.

Let me say it in a slightly more technical way.

Certain measurable quantities, Heisenberg discovered, are paired in an odd way. Two such pairs are momentum and location, and energy and time. Physicists call these odd couples *conjugate variables* or *incompatible observables*.

The more accurately we measure one member of the pair, the less accurately we're able to measure the other. Wonderland is wired so that we can never nail down both members with perfect accuracy.

Suppose we experiment on fast-moving helium nuclei, such as are found in cosmic rays. The better we measure a nucleus's momentum, the worse we're able to measure its position.

It's like a teeter-totter. The higher the accuracy of one measurable quantity, the lower the other one will be. There's no getting around it.

It's also like what videographers call a *rack focus*. When you have two subjects—one in the foreground and one in the deep background—and you focus on the front subject, the back subject immediately becomes blurry. And vice versa. You can never get both subjects into perfect focus.[12]

It's not that you're being sloppy or incompetent. It's how Wonderland rolls, for deep reasons we don't understand—and maybe never will.

PHYSICS IN WONDERLAND II

On September 26, 1905, *Annalen der Physik*—the same German journal that had published Einstein's paper on quantum physics three months earlier—published his theory of special relativity (SR).[13] On this remarkable day, the twenty-six-year-old phenom led physicists even deeper into the wilds of Wonderland.

Here are three shocking things that Einstein's theory claims about the world we call home. All of them are totally at odds with Aristotelian and Newtonian physics.

1. Wonderland has four dimensions.

Physicists since Aristotle believed the universe to have just three space dimensions: up/down, right/left, and forward/backward. These dimensions—labeled x, y, z—are easily described using ordinary 3D geometry.

Special relativity, however, claims that *time* is also a dimension, even though it behaves very differently than the three spatial dimensions. It's like saying men and women are both human, even though they behave differently.

Physicists call the four dimensions *spacetime* (one word) and label them x, y, z, t. Sometimes they can easily be described using ordinary 4D geometry—but not always.[14]

2. Certain superficial aspects of Wonderland are relative.

To physicists since Aristotle, *distances, times,* and *masses*—everyday quantities measured by rulers, clocks, and scales—appeared to be inflexible. It was expected that different observers would always agree on, say, the length of a foot, the duration of a minute, or the mass of a penny.

But SR claims that distances, times, and masses are *elastic*—like a rubber band. The length of a foot, the duration of a minute, the mass of a penny all depend on your particular situation. Specifically, on your speed relative to the thing you are measuring.

You aren't conscious of living in a rubbery Wonderland for one simple reason. The weirdness becomes obvious only at huge speeds.

Many people—especially those espousing a post-truth worldview—have pounced on this discovery and greatly distorted it. So before I go any further, let me make it crystal clear: The surprising revelations of special relativity notwithstanding, absolute truth *does* exist. For as you're about to see, even distance, time, and mass—any and all relative quantities—ultimately obey the *absolute* laws of physics.

Let me give you some concrete examples of what I mean.

First, find a hard plastic ruler and hold it in your hand. If it's well made, it'll be exactly twelve inches long.

Now suppose I drive past you at 60 mph with an identical twelve-inch ruler in hand. To me, your ruler will look *shorter* than twelve inches. Not by much, mind you. At 60 mph, I'll reckon your ruler to be 11.999999999999952 inches long.

Here's another example.

Use your smartphone's stopwatch app to measure out a minute. If your stopwatch is working properly, the minute will be exactly sixty seconds long.

Now imagine I zip past you at 60 mph with an identical stopwatch app. To me, your stopwatch will appear to run *slower* than mine. Compared to my minute, yours will seem 60.0000000000024 seconds long.

One final example.

Weigh a brand-new, shiny penny (which is 2.5 percent copper and 97.5 percent zinc). According to the US Mint, its mass is exactly 2.500 grams.[15]

Now, once again, suppose I drive by at 60 mph. To me, the mass of your penny appears *greater* than 2.500 grams—namely, 2.50000000000001 grams.

I can easily imagine you're shaking your head right now, protesting: "One and the same ruler can't have two different lengths! One and the same minute can't have two different durations! One and the same penny can't have two different masses!"

Actually, they can.

You just don't realize it.

"But, c'mon!" you say. "One person has to be wrong. You and I can't both be right! It's impossible!"

Welcome to Wonderland.

Welcome to the world you inhabit.

In our strange, strange world, the length of a foot, the duration of a minute, the mass of a penny—distance, time, and mass—are not absolute truths. They're optical illusions. Like the many facets of a single, hard diamond, they're superficial aspects of a deeper *absolute reality*.

3. Light has a sacred status in Wonderland.

We've already taken a look at this remarkable reality in chapter 4. Here I will give you a deeper understanding of it. But don't worry, I won't get too technical.

First, you must be mindful of this absolute truth: *Speed is relative*. Suppose you're driving on the freeway and a parked cop clocks you going 60 mph. To someone driving alongside you at 60 mph, you're standing still—in effect, you're going 0 mph because you're not moving relative to the other car.

So who's right? The parked cop or the motorist in the next lane? Both! Because speed is relative; it depends on your point of view or frame of reference.

This familiar fact shouldn't surprise you. Speed is made up of *distance* and *time*—it's defined as distance per time—both of which are relative quantities.

But now to the one shocking exception.

According to SR, the speed of light in a vacuum—299,792,458 meters per second—is *not* relative.[16] Let me explain.

Suppose a *light quantum* streaks across your room. Standing at the door with a souped-up radar gun, you clock it going, as expected, 299,792,458 meters per second. (Obviously I'm ignoring that your room is not a vacuum. But this simplification does not affect what I'm explaining.)

Now, how about to someone driving by at 1,000,000,000 mph? Here's the shocker: The light quantum will still seem to be traveling at 299,792,458 meters per second.

Unlike the speed of a car or anything else in the universe, the speed of light doesn't depend on one's point of view; it's the same for everyone, everywhere, always. It's an absolute truth, the only speed in the universe with that supreme status—for reasons, mind you, we do not understand. It's a mystery.

One final thing about light's puzzling sacredness: You and I can never accelerate to the speed of light, no matter how hard we try.

The harder we try, the more massive we become, making it all the harder for us to speed up. To paraphrase a popular television commercial, we can only cry out, "Help! I'm overweight, and I can't speed up!"

By the way, we physicists know that this phenomenon actually happens: Subatomic particles in atom smashers bloat in exactly this way as they accelerate toward the speed of light. We just don't understand the fundamental reason *why* the universe works this way.

PHYSICS IN WONDERLAND III

Think of gravity as a mysterious lover. You live with it every day, yet you know next to nothing about it.

Even Isaac Newton, who knew more about gravity than anyone else in his day, gave up and passed the buck to others: "Gravity must be caused by an agent acting constantly according to certain laws," he said, "but whether this agent be material or immaterial is a question I have left to the consideration of my readers."[17]

Centuries later, on November 25, 1915, Einstein claimed to crack the code. Standing before a physics class at the Royal Prussian Academy of Sciences in Berlin, he announced that gravity is caused by dimples in the elastic fabric of 4D spacetime.[18]

Yes, *dimples.*

The sun, for example, creates a humungous dimple the way a circus elephant would on a rubber trampoline. Earth produces a much smaller dimple. You and I make truly insignificant spacetime dimples.

Says Einstein: Whenever you travel across a spacetime dimple, you'll naturally follow its contours and, therefore, veer off course. (Having driven across many a pothole while living in Boston, I know the feeling.) Because the dimple is invisible, you'll understandably blame your altered course on an unseen force: *gravity*.

If you have the misfortune of traveling across a deep enough spacetime dimple, you'll fall into it and never come back out. We call those dimples *black holes*, the Bermuda Triangles of outer space.

To explain all this weirdness, Einstein needed to use a geometry far more exotic than Euclid's. For help, he hit up friends who were mathematical virtuosi—including Marcel Grossmann, a former college classmate.[19]

With their guidance, Einstein eventually found just the right non-Euclidean geometry for the job: a 4D geometry invented in the 1850s by the German prodigy Bernhard Riemann.

With Riemann's geometry, Einstein formulated a law of gravity that trumped Newton's old one—for me, the most beautiful equation in all of physics:

$$G_{\mu v} + \Lambda g_{\mu v} = \frac{8\pi G}{c^4} T_{\mu v}$$

You might hate math—lots of people do. But just for a moment, I invite you to see this equation as I do: as *a work of art*.

Its exquisite, SQ-inspired brushstrokes portray the strange beauty of a 4D Wonderland strewn with dimples the eye cannot see, IQ cannot prove, and the mind cannot fully imagine. Of a world—your world and mine—whose magical enchantments cannot be put into words.

Remember Joe, the split-brain patient we met in chapter 6?

When a photo of a frying pan was placed in front of Joe, he couldn't see it—or name it. The photo represented something invisible and beyond words.

But when Joe closed his eyes, he sketched a frying pan. It wasn't a conscious IQ that directed the pencil in hand. It was an inner awareness, a prescient SQ, that produced the drawing.

Einstein's equation is like Joe's sketch. It's the creation of an SQ that opens our eyes to truths we cannot verbalize.

Viewed this way, mathematics is more of an art form than a language. It's more like filmmaking, acting, music composing, sculpting, and dancing—all the SQ-powered talents that equip our species to perceive, process, and communicate extra-worldly truths and realities that leave us speechless.

Leonardo da Vinci, a hero of mine since boyhood, filled notebooks with beautiful drawings and beautiful equations that spoke of the natural and the supernatural. "Let no man who is not a Mathematician," he admonished, "read the elements of my work."[20]

For me, Leonardo represents the quintessential high-SQ-high-IQ scientist. Someone who, with his 3D glasses firmly in place, saw Wonderland in all its strange and stunning magnificence. According to an Italian miniseries on his life, Leonardo explained his stereoscopic perspective this way: "Since no intellect can penetrate nature and no language can explain its marvels, human thought is guided to the contemplation of the divine."[21]

BEYOND THE BUBBLE

The discoveries of modern physics are undeniably fascinating, but they're also far-out. Which might leave you wondering: *Who cares?*

Why should it matter to you whether the mass of a penny is fixed or elastic? Whether the universe has three dimensions or four (or twenty or more)? Whether gravity is an invisible force or the consequence of some cosmic pothole? None of this information will help you watch your weight, cope with your boss, or deal with the realities of sickness, aging, and death.

Perhaps it won't. But it does matter.

Years ago, actor-comedian Jim Carrey starred in a movie called *The Truman Show* about a thirtysomething man whose entire life has been broadcast live as a reality TV show—all unbeknownst to Truman. His everyday world appears to be utterly normal. But one day, he starts noticing certain oddities that lead him to make a shocking discovery: His entire everyday world is actually a soundstage—a small bubble of normalcy surrounded by a vast, jaw-dropping Wonderland populated by things and people he'd never seen nor could've ever imagined.

You are Truman.

Your everyday world is a bubble of normalcy. A bubble of mundane reality that on some days—perhaps, many—causes you to feel trapped, bored, even depressed.

Yet, if you know and understand the discoveries of modern physics, they are liberating. Thanks to these revelations, you now realize—like Truman—that there is more to life than your everyday bubble of normalcy.

You now realize that you are an integral part of a Wonderland that exists—not *out there* somewhere but *all around* and even *within* you—within the very cells of your body and brain. A dazzling, translogical Wonderland not even Lewis Carroll could've ever imagined.

10

HAVING FAITH IN ASTRONOMY

Everything you've learned in school as "obvious" becomes less and less obvious as you begin to study the universe.

R. BUCKMINSTER FULLER

Growing up in East Los Angeles, I lived under a night sky awash in light pollution. The first time I ever saw a truly dark sky was when my high school buddies and I camped out in the Mojave Desert.

Wow!

The midnight sky was pitch black, and the stars—like flashing gemstones—seemed close enough to reach up and grab. The Milky Way looked like an incandescent river.

A truly clear night sky doesn't just bowl you over with its cosmic majesty; it makes you wonder what might be out there. It stirs something deep in the human soul.

It's the reason astronomy is arguably the most emotional and spiritual of all the sciences.

"For every one, as I think, must see that astronomy compels the soul to look upwards and leads us from this world to another," writes Plato in his *Republic*.[1]

Astronomy is certainly an ancient science. Civilizations as far back as Mesopotamia circa 3500 BC practiced astronomy—alongside its notorious cousin, astrology. Ancient priests were the original astronomers, constantly scrutinizing the heavens for portentous signs of religious, agricultural, and social importance.

Astronomy is also a peculiar science; you might say it's not even a true science at all. Unlike the hard sciences—such as physics, chemistry, or biology—astronomy seeks to understand objects that are completely beyond our reach and control. You can't make stars and planets do your bidding as you can pulleys or atoms or living cells. So you can't easily do controlled experiments, the bread and butter of any credible science.

Most of the time, all you can do is analyze the *light* from faraway objects, deducing from it as much information as humanly possible.

"Light brings us the news of the Universe," declared British scientist Sir William Bragg. "Coming to us from the sun and the stars, it tells us of their existence, their positions, their movements, their constitutions, and many other matters of interest."[2]

But the intelligence that can be squeezed from this heavenly light is limited. Especially since we now have evidence that 95 percent of the universe is imperceptible.

That's right, it appears that *95 percent* of the cosmos is hidden from us—in the form of dark matter, dark energy, and other covert phenomena. It emits no detectable light whatsoever. It is *invisible*.[3]

That means astronomers are forced to operate mostly in the dark. *Literally.* They must rely on faith—ideally IQ-and-SQ-based enlightened faith—to believe and defend their conjectures about our mostly invisible, otherworldly cosmos.

"Dark matter isn't supernatural," says British astrophysicist Richard Massey of Durham University, "but its mysterious behavior certainly brings that idea to mind."[4]

The same is true about dark energy. It's our name for the invisible *something* that seems to be causing the universe to swell at an ever-faster speed. But we have absolutely no idea what it is.

"Dark energy is something even more bizarre than dark matter," remarks American astrophysicist Saul Perlmutter.[5] In 1998, Perlmutter and two colleagues discovered the universe's accelerating expansion, earning them the 2011 Nobel Prize in physics.

What is dark matter? What is dark energy? They aren't the only deep mysteries astronomers are wrestling with. In this chapter, I will describe for you four of the most baffling ones: How big is the universe? How old is the universe? How did the universe begin? Does extraterrestrial life exist?

This last mystery belongs to the subdiscipline of *planetary astronomy*, which studies how planets form and whether Earth is the only planet with life.

The other three mysteries belong to the subdiscipline of *cosmology*. It's the one I fell in love with in grad school. Cosmologists focus on the big picture, on what the whole universe looks like and how it got to be that way.

As you're about to see, cosmology taxes the scientific method and human mind *beyond their limits*. Cosmology operates in a realm far beyond conventional, controlled experiments; far beyond what we can see or even imagine; far beyond logic and IQ. A realm accessible only to our unsighted, yet penetrating, SQ.

For my money, it doesn't get more exciting than that.

"While this discipline [physical cosmology] is a convolution of elementary particle theory, general relativity, and astronomical observations, there is still room for elements of mysticism and imagination," remarks University of Oregon physicist and cosmologist Gregory Bothun. The mysteries of cosmology are so profound, he says, that "there remains no clear and preferred model for the origin and evolution of structure in the Universe."[6]

HOW BIG IS THE UNIVERSE?

Cosmologists of old disagreed strongly about the size of the universe. Some said it was infinitely large. Others said it had limits.

Today we have powerful enough telescopes and clever enough methods to size up the universe with some reliability. One of these methods is called the *cosmic distance ladder*, which uses some clever techniques to estimate distances, step-by-step like the rungs of a ladder, far out into the cosmos.

The cosmic distance ladder is a glowing tribute to my fellow cosmologists' ingenuity and resourcefulness. It makes enormously smart use out of what little information we're able to glean from heaven's light.

But clever as the ladder is, each of its techniques is not entirely reliable. Each distance-reckoning technique—each rung—depends on the accuracies of all the rungs before it. So if any technique fails, it undermines everything after it. To paraphrase an old saying, the cosmic distance ladder is only as strong as its weakest rung.

Here is a brief summary of the ladder's main rungs. (For a fuller explanation, please see Appendix A: The Cosmic Distance Ladder). For a sense of scale, bear in mind that one light-year (ly) equals nearly *six trillion miles.*

COSMIC DISTANCE LADDER

METHOD USING ...	RELIABLE OUT TO ...
laser, radar, radio	solar system 600,000,000 mi
parallax	nearby stars 30,000 ly

METHOD USING ...	RELIABLE OUT TO ...
ordinary starlight	Milky Way 200,000 ly
exotic starlight	nearby galaxies 10,000,000 ly
galactic spin, supernovae, redshifts	deep space 10,000,000,000 ly

Based on the ladder's best estimates, we now believe the observable universe to be about 92 billion light years across. That's 5.5 x 10^{23} or 550,000,000,000,000,000,000,000 miles across.

The *actual* physical universe is much larger than that, but its outermost regions are moving away from us so fast that its light will never reach us. Those regions will always remain hidden from us—like a cosmic Road Runner forever beyond the reach of Wile E. Coyote.

Without being able to see the outermost regions of the universe, we can't possibly use the cosmic distance ladder to reckon its actual, ultimate size. But thanks to general relativity, we can make an educated guess based on the numerical value of the observable universe's *critical density*. If you've ever been on a diet, think of critical density as your *target weight.*

If the universe is overweight—if its total mass-energy density *exceeds* the critical density—then the size of the universe is *finite.*[7] We cosmologists call this a *closed* universe. One day, owing to its obesity, such a universe will collapse on itself.

If the universe is at or under its target weight—if its total mass-energy density is *equal to or less than* the critical density—then its size is *infinite.* We call such a universe *flat* or *open*, respectively. In both cases, such a universe keeps expanding, diluting, and transforming into a cold, dead nothingness.[8]

Improbable as it is, our universe appears to have precisely the target weight. It's not too fat and not too skinny. It's just right.

This means the universe is *infinitely large* and *flat*, albeit barely so.[9] In our universe, in other words, venturing out into space is like traversing a vast wilderness whose horizons we can never reach. Ever.

But there's one wrinkle.

Recent data from the Planck Orbiting Observatory implies that the universe is actually *obese*; that it has a density far *greater* than the critical density. If so, our universe is actually *finite* and *closed*. In that case, venturing into space is like traveling in circles: After a while, you return to where you started.[10]

HOW OLD IS THE UNIVERSE?

For many centuries, no one ever asked this question. Everyone agreed the cosmos was *ageless* and *static*. That what we see today is what has always existed and always will exist.

Astonishingly, that worldview prevailed until the early twentieth century. That's when scientists made a pair of utterly devastating discoveries.

In 1915, young Albert Einstein published his theory of general relativity, cocksure it would affirm the ageless-static theory (AST). *But it didn't!* Instead—to his horror—it raised the possibility that the universe is expanding.

Rattled, Einstein quickly amended his theory's beautiful equation of gravity, the one I showed you in chapter 9. He inserted a fudge factor called *lambda*, the cosmological constant.

With lambda in place, Einstein's equation squared perfectly with the AST. All was well in the world of cosmology.

But then in 1929, American astronomer Edwin Hubble published data showing that galaxies are moving away from each other, like so many pieces of shrapnel after an explosion. He summarized the results in what is now called Hubble's Law:

$$v = H \times d$$

Translated into English, it says:

v (the speed at which a galaxy is moving away from us)

equals

H (the Hubble constant) *times* **d** (the galaxy's distance from us)

In short, the *farther* away the galaxy, the *faster* it is moving away from us.

Once again, the world of cosmology was thrown into chaos.

But this time cosmologists could not avoid the inevitable, shocking conclusion: Our universe appears to have exploded into being and is now expanding. It was just as Einstein's original, unadulterated equation of gravity had indicated.

Cosmologists deep-sixed the venerable AST and for the first time in human history asked themselves the question: *How old is the universe?*

They quickly recognized that Hubble's constant gave them the answer. Here's what I mean:

- **Big H** (fast expansion) means the universe reached its current size quickly. Which means it's relatively young.
- **Small H** (slow expansion) means the universe has taken a long time to balloon out to its current size. Which means it's relatively old.

All cosmologists needed to do, then, was pin down the value of H. But that proved to be easier said than done.

The value of H, you see, depends on estimates of cosmic distances.

But as I've already explained, those are not precise, even with the help of the sharp-eyed Hubble Space Telescope.[11]

Also, the value of H is not constant; it changes with time. Which means cosmologists need to know what the universe looked like at every stage of its growth, starting from when it was a newborn.

But how can you possibly get *baby pictures* of the universe? Answer: by peering far enough out into space.

Light from very distant objects takes billions of years to reach our telescopes; so its arrival conveys images of the infant cosmos. The cosmos of billions of years ago.

In 2013, astronomers gathered all their best data and concluded that H = 70. Which means the universe is 13.8 billion years old.

But even as I write this, various cosmologists are questioning the value of H.

In 2019, an American group—using the Hubble Space Telescope *and* the farthest rungs of the cosmic distance ladder—published a paper claiming that H = 76. That would mean the universe is a bit less than 13 billion years old.[12]

More recently, a European group published evidence that H = 82.4, which translates to an age of only 11.4 billion years.[13]

The haggling is not likely to end soon.

"The discrepancy," remarks Adam Riess, Nobel Prize–winning astronomer at the Space Telescope Science Institute, in Baltimore, "suggests that there's something in the cosmological model that we're not understanding right."[14]

HOW DID THE UNIVERSE BEGIN?

In the beginning, *there was no beginning*. That's because, in the beginning, astronomers believed in the AST. It wasn't until the early 1930s that cosmologists conceded their mistake and embraced the idea of an expanding universe.

A few years before that happened, however, an obscure Belgian cosmologist and Roman Catholic monsignor named Georges Henri Joseph Édouard Lemaître jumped the gun—and was severely ridiculed for his brashness.

Lemaître was no slouch. He had *two* PhDs—one in mathematics from the Catholic University of Louvain and one in physics from MIT.

In 1927—two years before Hubble's bombshell discovery about the expanding universe—Lemaître published a paper with the imposing title "A Homogeneous Universe of Constant Mass and Growing Radius Accounting for the Radial Velocity of Extragalactic Nebulae." In it, he made the astonishing claim that the universe hatched from a "cosmic egg" or "primeval atom"—a foreshadowing of today's big bang theory.

In October 1927, Lemaître traveled to Brussels for the Fifth Solvay Conference, where he buttonholed Albert Einstein and explained his heretical idea. "Your calculations are correct," Einstein reportedly sneered, "but your grasp of physics is abominable."[15]

Ouch.

Worse, even after Lemaître's idea was vindicated by Hubble's discovery, cosmologists still turned up their noses. "Philosophically," British astronomer Sir Arthur Stanley Eddington said, "the notion of a beginning of the present order of Nature is repugnant to me. . . . I see no way round it; but . . . I should like to find a genuine loophole."[16]

In 1949, during a BBC radio broadcast, British astronomer Fred Hoyle argued stubbornly in favor of the AST. He denounced Lemaître's idea, describing it as "the hypothesis that all the matter in the universe was created in one big bang at a particular time in the remote past."[17]

Ironically, Hoyle's sarcastic term "big bang" stuck—but not in the way he intended. Today the big bang theory (the standard cosmological model, or SCM) is sacred scientific dogma.

But even though we've come a long way since Einstein dissed poor Lemaître, cosmology still has a long way to go. For as we saw in chapter 2, the SCM has serious problems.

They might be solved one day; but for now, cosmologists are still wondering, *How did the universe begin?*

DOES EXTRATERRESTRIAL LIFE EXIST?

Okay, so you stare up at the night sky, marvel at the vastness of space, and think to yourself, *There's gotta be somebody out there! But if so . . .* where is everybody?

In 1950, the Nobel Prize–winning physicist Enrico Fermi famously asked that very thing: "Where are they?" Where in the universe are all the little green men (LGM)?

Astronomers estimate there are roughly 100–200 billion galaxies in the universe and about 100–400 billion suns in our Milky Way galaxy alone. So it seems reasonable to suppose there's a sun out there somewhere being circled by a planet with life on it.[18]

But we haven't found one, and no little green man or woman has ever knocked on anyone's door. Why not? This head-scratcher is called the *Fermi Paradox.*

Astronomers have been formally searching for LGM since 1960, when Frank Drake used the world's largest radio telescope, in Arecibo, Puerto Rico, to listen for signals from LGM. He heard *nothing* of the sort.*

For the past two decades, planetary astronomers have peered long and hard at the heavens in search of extraterrestrial planets, or what we call *exoplanets.* Using the world's strongest telescopes, including the powerful Hubble Space Telescope and Kepler spacecraft, they've found evidence for more than 4,300 other worlds.

Now, please understand that most exoplanets are way too far away for astronomers to actually see. Typically, we infer their presence by the wobbles they supposedly cause in the orbits of their host stars

*Tragically, in December 2020, the Arecibo radio telescope collapsed and is now permanently out of commission.

and/or by the shadows they supposedly cause as they pass in front of the stars.

From those inferences, astronomers estimate an exoplanet's period of rotation (its solar year) as well as its diameter and distance from its host star. From this information, we can tell if the exoplanet falls into the so-called *Goldilocks zone*. That is, if its properties are "just right" to support life.[19]

Of all the exoplanets astronomers believe they've found, take a guess at how many could have LGM living on them? The correct answer is zero, zilch, *nada*.

And that's according to NASA. In response to a question on their official Exoplanet Exploration website—"Are there any exoplanets like Earth?"—NASA replies, "We have found many Earth-sized rocky exoplanets, some of which are in the habitable zones of their stars," but "we haven't found a planet that can support life like Earth. So far, our home is unique in the universe."[20]

Think about that.

As far as science now knows, our solar system is *unique*. Earth is *unique*. You and I are *unique*—in the whole of the universe!

What are the odds that one day we'll discover LGM?

I first became seriously interested in exobiology—the study of extraterrestrial life-forms—as a grad student at Cornell. I had the privilege of being taught by Carl Sagan and Frank Drake, the co-fathers of the search for extraterrestrial intelligence (SETI). In fact, Frank was a member of my PhD thesis committee.

One of the chief lessons I learned from them is that there are at least two reasons why it's nearly impossible to answer the question of whether LGM exist.

First, it's not easy to know what constitutes life.

Every living thing on Earth is constructed from six chemical elements: carbon, hydrogen, nitrogen, oxygen, phosphorus, and sulfur (CHNOPS). But there are *ninety-four* natural elements plus *twenty-four* synthetic ones.[21] Is it possible a completely unknown form of life could be created from elements other than CHNOPS?

The short answer is *we don't know.*

Exobiologists and plenty of sci-fi writers have envisioned life-forms made of silicon, boron, or even germanium. Some imagine life-forms based not on water but ammonia. Still others speculate about *non-biological* life-forms made of metal, plasma, or pure consciousness.

Because we don't know, the speculations are wild-eyed and endless.

Second, we don't even know how Earth's CHNOPS-based life began. The standard evolutionary theory (SET) only purports to explain what happens to life *after* it begins, not how it actually begins.

Evolutionary biologists speculate it began on its own, a conjecture known as *abiogenesis.* Generally speaking, their ideas fall into two broad categories, both of which sound like science fiction.

I call the first category *It Came from Outer Space.*

In this scenario, the essential ingredients of life—water, amino acids, and nucleotides—are imagined to have rained down on Earth from some external agent—such as a meteorite, comet, or space alien. Richard Dawkins, evolutionary biologist emeritus at the University of Oxford, explains it this way:

> Nobody knows how it [life on Earth] got started. . . . It could be that at some earlier time, somewhere in the universe, a civilization evolved by probably some kind of Darwinian means to a very, very high level of technology, and designed a form of life that they seeded onto perhaps this planet. . . . And that designer could well be a higher intelligence from elsewhere in the universe.[22]

The *It Came from Outer Space* scenario is corroborated by our discovery that comets and meteorites contain organic molecules necessary for life—even amino acids, the building blocks of proteins.

For instance, Meteorite 2008 TC3, which pelted northern Sudan in 2008, contains nineteen different amino acids. And Comet 67P/Churyumov-Gerasimenko contains many different organic molecules and glycine, the simplest of all amino acids used by life on Earth.

I call the second category *It Came from the Black Lagoon.*

In this scenario, the essential ingredients of life were pieced together by some chemical process right here on terra firma. This was Charles Darwin's pick.

In 1871, Darwin speculated wistfully that Earth's very first protein macromolecules might've been cooked up by accident "in some warm little pond with all sorts of ammonia & phosphoric salts—[with] light, heat, electricity [etc.] present."[23]

In 1953, American chemists Harold Urey and Stanley Miller put Darwin's dreamy scenario to the test. They zapped a mixture of methane, ammonia, and hydrogen gases with electrical sparks and produced amino acids and other organic molecules.

The results were hailed at first but then questioned for various technical reasons.

One problem is that Earth's primitive, prebiotic atmosphere is believed to have been mostly carbon dioxide, nitrogen, and water vapor, *not* methane, ammonia, and hydrogen. One of Miller's students later corrected the gas mixture and repeated the landmark experiment, but the results are still contentious.

Another problem is that amino acids with identical chemical sequences come in two geometrical varieties: right-handed and left-handed. Life on Earth uses *only* the latter kind. So if an experiment wishes to claim it has successfully simulated the process that assembled life on Earth, it must produce *only* left-handed amino acids.

The Urey-Miller experiment does not do this. It produces the two kinds of amino acids equally, meaning that it does *not* represent the extraordinary, life-generating process that selected out the left-handed molecules and then assembled them into the organisms we see today.

Canadian scientists at McMaster University in Ontario recently completed a more souped-up version of the Urey-Miller device, which they call a *planet simulator*. By varying the temperature, humidity, pressure, atmosphere, and radiation levels inside a microwave oven-sized terrarium, the scientists can simulate the environment of early Earth or any other planet. It remains to be seen whether it will produce any reliable, replicable, and relevant information concerning the Black Lagoon scenario.

Even if we ever do find the right kinds of amino acids in meteorites and comets or create them in the laboratory, we will still be a long, long, long way from finding or creating actual life. I don't have the space here to fully explain what I mean, but here's a taste.

Yes, amino acids are the "building blocks" of life. But it takes scores of them hooking up just right. And they must also fold themselves into just the right shape in order to create a single, healthy protein.

This intricate, exquisite process is like origami. Just one wrong fold and the end product won't look right or function properly.

For example, a single hemoglobin protein is made of 574 amino acids bound together in a three-dimensional, precisely interwoven, origami-like knot. A single mistake in the complex creation process produces a dysfunctional protein—such as we see in people with sickle-cell anemia, some of whom don't survive childhood or even infancy.[24]

A living organism relies on *hundreds* of such carefully constructed proteins to function properly. Even *Mycoplasma genitalium*, a sexually transmitted pathogen with one of the tiniest, most primitive genomes on the planet, relies on upwards of six hundred different, perfectly constructed proteins. You and I require anywhere from ten thousand to several billion different, intricately built proteins.[25]

Even then, proteins are not life-forms. They cannot replicate themselves, an essential feature of anything claiming to be alive.

For that, you need something like an RNA molecule, or ribonucleic acid—for which you need nucleotides, building blocks infinitely more complex than amino acids or proteins. As of this writing, nucleotides have not been discovered in space, nor is there irrefutable evidence that they've been created in the lab unassisted and in an environment that simulates a primordial Earth.†

Even if we ever did find or make nucleotides, we would still be a long, long, long way from finding or creating actual life. For, in addition to the material ingredients for life, we need one more thing: *information*.

In other words, we would need a recipe *and* a chef to mix the ingredients into something other than a nondescript blob. Without the guiding hand of information, it's very unlikely that Earth's 4.5-billion-year lifespan would be anywhere near enough time for CHNOPS to assemble *themselves* into amino acids, proteins, and nucleotides—and then for those amino acids, proteins, and nucleotides to accidentally bake themselves into exquisite specimens like you and me.

"Biochemical systems are exceedingly complex, so much so that the chance of their being formed through random shufflings of simple organic molecules is exceedingly minute, to a point indeed where it is insensibly different from zero,"[26] explains British astronomer and Atheist Sir Fred Hoyle, whom I had the pleasure of getting to know. "For life to have originated on the Earth it would be necessary that quite explicit instructions should have been provided for its assembly."[27]

"The probability that at ordinary temperatures a macroscopic number of molecules is assembled to give rise to the highly ordered structures and to the coordinated functions characterizing living organisms is vanishingly small,"[28] concurs the Russian-born Belgian

†Do not confuse nucleotides with nucleobases. A DNA molecule is shaped like a spiral staircase. A nucleobase is a rung of the staircase. A nucleotide is more complex, as it includes a rung *and* a section of the outer structure of the staircase.

physical chemist and Nobel laureate Ilya Prigogine, whom I also had the pleasure of knowing. "The idea of spontaneous genesis of life in its present form is therefore highly improbable, even on the scale of the billions of years during which prebiotic evolution occurred."[29]

What, then, are the odds that LGM exist?

Allow me to introduce you to the Drake Equation, named after my former professor Frank Drake. It gives a rough estimate of how many intelligent civilizations are likely to exist *in the Milky Way galaxy alone*.

The equation takes into account seven key factors:

R_* = How frequently are suns born whose light could conceivably sustain intelligent life?

f_p = What fraction of those stars have planets?

n_e = How many of those planets, per solar system, have environments suitable for life?

f_l = What fraction of those planets actually host life?

f_i = What fraction of those life-bearing planets have *intelligent* life?

f_c = What fraction of those intelligent civilizations broadcast detectable signals into space?

L = How long do those civilizations broadcast detectable signals into space?

Drake and his colleagues calculated that our galaxy alone should house about 1,000–100,000,000 intelligent civilizations.[30]

This means that about *100 trillion–20 quintillion* intelligent civilizations should exist in the entire, visible universe. A veritable explosion of technologically advanced LGM.

And yet, notwithstanding today's excitement about possible life on Mars and the discovery of exoplanets, we're still living with Fermi's Paradox. When we look up or listen into deep space with our finest, most exquisite instruments, we find no hard evidence for LGM and hear only crickets.

Why?

According to a team of researchers at Oxford University's Future of Humanity Institute, it's because we've been assigning overly optimistic numbers to the Drake Equation.[31] We want so badly for there to be LGM, we've grossly overestimated the number of civilizations that might exist out there.

When the Oxford folks assigned realistic values to the seven factors—based on an honest evaluation of the uncertainties that afflict our very best chemical, biological, physical, and astronomical knowledge—Drake's famous equation predicts far, far fewer than 1,000–100,000,000 intelligent civilizations per galaxy. The median number plummets to something as low as 0.0000000000000000000 0000000000000008 (that's an eight preceded by thirty-four zeroes).

In a paper submitted to the *Proceedings of the Royal Society of London*, the authors concluded, "We find a substantial probability that we are alone in our galaxy, and perhaps even in our observable universe."[32] If any LGM do exist out there somewhere, the authors add, they're somewhere over the rainbow—"quite possibly beyond the cosmological horizon and forever unreachable."[33]

Remarkably, the Bible agrees. As we'll discover in the next chapter, sentient beings do exist "beyond the cosmological horizon."

One of them in particular visited Earth two thousand years ago, and his stay is documented in striking detail in the most widely read book in human history—an ancient tome that has survived centuries of scrutiny by countless skeptics and is today supported by volumes of well-documented historical and physical evidence. A book that squarely takes on the question: "Are we alone?" and gives us the definitive answer: *No, we are not.*

11

HAVING FAITH IN GOD

"I AM THE LIVING GOD, The Resurrection and The Life; whoever trusts in me, even if he dies, he shall live."

JOHN 11:25-26, ANT

After Laurel and I finished reading the Bible for the first time, we began listening to a weekly radio program called *Voice of Prophecy*, hosted by Pastor Lonnie Melashenko. It was our introduction to mainstream Christianity—or so we thought.

Intrigued enough by what we were hearing, we sent for the program's Bible study guides. Their perspectives on Scripture, we found, were fairly consistent with our own.

The final lesson, however, was not. It analyzed the history of calendars and then argued that Saturday, not Sunday, was the true Sabbath. Also—and this was the kicker—it implied that anyone observing the Sabbath on Sunday was in danger of going to hell.

As novice explorers of the Christian worldview, Laurel and I were gobsmacked!

Our eternal fate depended on which day of the weekend we chose to rest? Really?

Curious to find out more about the people behind it all, I did some digging (this was before search engines) and discovered they were Seventh-day Adventists, a Christian denomination with roughly 20 million adherents worldwide.[1] They're a tiny fraction of the total two billion–plus Christians in the world, but they are well known for being a friendly lot and having an impressive multimedia outreach ministry.

This jarring experience left Laurel and me wondering what to do next. We wanted to learn more about Christianity, but the question now was: *Whose* Christianity?

The more we looked into it, the murkier things became. We learned there were disagreements even among fundamentalist Christians who believe in the infallibility of the Bible and a literal interpretation of the text. Some of the disputes are over seemingly small things; for example, over what qualifies as a proper baptism. Some Christians believe a sprinkle of water is sufficient; for others, only full immersion counts with God.

It reminds me of a poem I ran across by William Blake. Here's the opening stanza:

The vision of Christ that thou dost see
Is my vision's greatest enemy.
Thine has a great hook nose like thine;
Mine has a snub nose like to mine.
Thine is the Friend of all Mankind;
Mine speaks in parables to the blind.
Thine loves the same world that mine hates;
Thy heaven doors are my hell gates.
Socrates taught what Meletus
Loath'd as a nation's bitterest curse,
And Caiaphas was in his own mind
A benefactor to mankind.
Both read the Bible day and night,
But thou read'st black where I read white.[2]

Christians also disagree about the most accurate translation of the Bible. BibleGateway.com, a popular online reference, offers access to *sixty-one* different translations in English alone. According to the American Bible Society, the total number of distinct English translations of the Bible is about *nine hundred.*[3]

Some of these translations are necessary to keep up with the evolution of our language. We don't speak the same English today as they did in the early seventeenth century, when the King James Version (KJV) was first published.

Here's how Job 36:32 reads in the KJV: "With clouds he covereth the light; and commandeth it not to shine by the cloud that cometh betwixt." Now here's the modern-English New Living Translation of the same verse: "He fills his hands with lightning bolts and hurls each at its target."

Scholars use different strategies to translate from the original Hebrew, Greek, and Aramaic. While some use word-for-word translations, others use thought-for-thought translations.

Faced with all these choices, Laurel and I felt like kids with a nickel in a candy store. But we didn't let the overwhelming confusion stop us; we pushed ahead.

In particular, I continued pursuing my interest in the similarity I'd noticed between the New Testament (NT) and quantum physics. I recognized that the NT, like quantum physics, was quite possibly espousing profound truths and realities about the universe. Not obvious, trivial, logical truths and realities but baffling, deep, translogical ones. Truths and realities that cannot be seen, proven, or even imagined.

This demanded of me—of anyone diligently seeking truth—an extra measure of open-mindedness, perspicacity, and patience.

I knew plenty of people who for various reasons had shrugged off the Bible; most of them had never even read it. *I had been one of them!*

Some people faulted the Bible for excluding certain apocryphal books, all of which I'd now read and learned were rejected for many important reasons.[4] Other people argued endlessly and fruitlessly about the best way to translate this word or that verse in the Bible.

I knew such people would always exist. And that Christians would always disagree about which interpretation of Scripture or which translation of the Bible is best.

But none of this noise put me off. In fact, it only *added* to my intuition about the similarity between the NT and quantum physics.

I say that because we physicists disagree strongly and endlessly about how to interpret quantum physics. The orthodox version of QP is called the Copenhagen interpretation; but there are countless other interpretations. Think of them as competing denominations—including the Many Worlds interpretation, the De Broglie–Bohm interpretation, and the transactional interpretation.[5]

So I knew full well that the NT could not be shrugged off simply because there are disagreements about how to interpret it. And there was another, huge, reason I couldn't shrug off the NT: I had already discovered that, despite its apparent strangeness, the Christian worldview, like quantum physics, was consistent with the best available evidence.

The evidence didn't amount to *proof*—but then, no such thing exists, as I've already explained. Quantum physics isn't provable, either, and it never will be, for the same reasons.[6]

In fact, Albert Einstein, notably, remained skeptical about quantum physics to his dying day. "Quantum mechanics is very impressive," he said. "But an inner voice tells me that it is not yet the real thing."[7]

Notwithstanding all the noisy skeptics and impassioned disagreements between even Christian scholars, I knew that when evaluating the Christian worldview, I needed to at least respect it. That is, I needed to probe it forcefully—as I'd done with other religions—but also open-mindedly, without prejudice.

If the Christian worldview held up to my scientific scrutiny, I was confident I'd be able to see past the haze of competing interpretations and discern not only Christianity's more obvious (but no less important) absolute trivial truths but its essential, inscrutable, translogical truths as well.

I liken it to the experience of listening to music. I've attended many performances of Beethoven's symphonies, and some of them were quite bad. But in every case, Beethoven's inexpressible genius always stood out.

In order to judge the heart and soul of the New Testament, I knew I had to focus my hard-nosed examination on its central figure: Jesus Christ. This meant taking on mainstream Christianity's most audacious claims about him, namely:

- Jesus is the human incarnation of the God described in the Bible, who created the universe and authored life.
- Jesus actually said and did what the NT claims he said and did, which include prophecies and miracles.
- Jesus is the Messiah foretold by ancient Hebrew prophecies.
- Jesus was born to a virgin.
- Jesus lived a sinless life.
- Jesus was crucified for a cosmic purpose: to repair the breach between humanity and God, which is produced and maintained by our constantly rebellious behavior.
- Jesus came back to life following his crucifixion.
- Jesus is part of a Trinity that includes God and something called the Holy Spirit.
- If you genuinely *believe in* Jesus—not just believe that he exists but sincerely repent and accept his sacrifice for you, thus making peace with God—you will coexist harmoniously with God forever, no matter who you are or what you've done in the past.

Reporting the complete results of my years-long investigation of Christianity would take an entire book of its own. What follows are some of the important highlights.

PROPHECIES

The Bible claims to be more than just a historical document. It claims to outline the story of our species, *Homo sapiens sapiens*, for all time.

In broad strokes, the Bible purports to explain how you and I and the universe came to be; the root cause of our bad behavior and today's unkind, unjust, and despoiled world; and the bleak-bright future we and Earth are hurtling toward.

In doing this, the Bible reveals something quite remarkable about Christianity: its belief in *linear time*.

This is no small thing. In my book *Amazing Truths: How Science and the Bible Agree*, I explain that other religions—for example, Hinduism and Buddhism—believe in circular time. Believe it or not, long before Christianity appeared on the scene, so did *science*.[8]

What's more, the sacred literature of other religions tends to be atemporal; that is, it describes people and events in a mostly timeless setting.

By sharp contrast, the Bible is diligent about specifying times and places for its people and events. In *Amazing Truths*, I cite many examples of this.

What's more, the Bible's historic authenticity is corroborated by a great deal of archaeological evidence. As of June 2017, according to Lawrence Mykytiuk, a renowned expert in Hebrew and Semitic Studies, science has confirmed the existence and time stamps of at least fifty-three people in the Old Testament alone; and the number continues to increase.

"Their names appear in inscriptions written during the period described by the Bible," explains Mykytiuk, "and in most instances during or quite close to the lifetime of the person identified."[9]

The Bible also makes predictions about the future. Many of these prophecies are about specific events at specific times—which, again, is no small thing.

In science, we believe that a hypothesis is credible only if it makes predictions that can be tested and possibly refuted. Claiming the moon is made of green cheese is legit because you can go to the moon and check it out. But someone claiming they dreamt of hitting the lottery is not because there's no way you can test the assertion.

Many religious claims—including Christian ones—are not legitimate scientific hypotheses because they can't be independently, objectively tested or possibly refuted. It doesn't mean they're not true; it only means they're not strictly scientific in nature.

This proviso applies to science as well.

The claim that multiple universes exist; or that science can explain everything; or that science and technology have more good outcomes than bad ones—these all might be true. But because they cannot be tested and possibly refuted, they do not qualify as legitimate scientific hypotheses. Instead, they fall under the category of *religious and philosophical beliefs*.

Among the hundreds of Old Testament prophecies are ones that foretell the coming of a Messiah.[10] More precisely, the OT appears to predict *two* Messiahs: one that suffers and one that triumphs. Or, quite possibly, a single Messiah who both suffers and triumphs.[11]

Mainstream Christians believe in the latter possibility and that Jesus is that suffering-triumphant Messiah. In fact, they believe he *had* to suffer in order to triumph—and that he'll make a second, permanent appearance on Earth sometime in the future.

Many of the Old Testament messianic prophecies are vague and can be defended persuasively, I think, only with the benefit of hindsight. But others are more like scientific hypotheses: They can be tested and possibly refuted. Those are the ones that caught my eye.

Here are some examples. Scholars affirm that these prophecies were written between BC 470 and BC 735—roughly 475 to 740 years *before* the birth of Jesus.[12]

> All right then, the Lord himself will give you the sign. Look! The virgin will conceive a child! She will give birth to a son and will call him Immanuel (which means "God is with us").[13]

> But you, O Bethlehem Ephrathah, are only a small village among all the people of Judah. Yet a ruler of Israel, whose origins are in the distant past, will come from you on my behalf.[14]

> Rejoice, O people of Zion! Shout in triumph, O people of Jerusalem! Look, your king is coming to you. He is righteous and victorious, yet he is humble, riding on a donkey—riding on a donkey's colt.[15]

> They [the family of David and the people of Jerusalem] will look on me whom they have pierced and mourn for him as for an only son. They will grieve bitterly for him as for a firstborn son who has died.[16]

> He was despised and rejected—a man of sorrows, acquainted with deepest grief. We turned our backs on him and looked the other way. . . . And we thought his troubles were a punishment from God, a punishment for his own sins! But he was pierced for our rebellion, crushed for our sins. He was beaten so we could be whole. He was whipped so we could be healed. All of us, like sheep, have strayed away. We

> have left God's paths to follow our own. Yet the LORD laid on him the sins of us all. . . . He was led like a lamb to the slaughter. . . . Unjustly condemned, he was led away. . . . He had done no wrong and had never deceived anyone. But he was buried like a criminal; he was put in a rich man's grave.[17]

> After this period of sixty-two sets of seven, the Anointed One will be killed, appearing to have accomplished nothing, and a ruler will arise whose armies will destroy the city [Jerusalem] and the Temple.[18]

> As my vision continued that night, I [Daniel] saw someone like a son of man coming with the clouds of heaven. He approached the Ancient One and was led into his presence. He was given authority, honor, and sovereignty over all the nations of the world, so that people of every race and nation and language would obey him. His rule is eternal—it will never end. His kingdom will never be destroyed."[19]

> "The LORD will mediate between nations and will settle international disputes. They will hammer their swords into plowshares and their spears into pruning hooks. Nation will no longer fight against nation, nor train for war anymore."[20]

These Old Testament prophecies, and others I haven't itemized here, jibe with the NT accounts of Jesus' time on Earth. Does that prove Jesus is the prophesied Messiah? *No.* Is the mainstream Christian interpretation of these prophecies the only possible one? *No.*

But, I repeat, *proof* in matters as complex as this isn't an option. The best we can ever say about any hypothesis—scientific or otherwise—is that it's consistent with the best available *evidence.*

That's what you can say here.

The hypothesis that Jesus is the prophesied Messiah is consistent with the best available evidence. That evidence, as we're about to see, includes not only the NT reports but other corroborating, extra-biblical sources, both Christian and non-Christian.

THE BIRTH OF JESUS

The NT's account of Jesus' birth fascinated me immediately because of the astronomical event—the Star of Bethlehem—that supposedly heralded the event. I won't get into it here, but I'm presently working with others on a project that explores the subject in a fresh, engaging way.

I was also intrigued by the claim that Jesus was born of a virgin. The angel Gabriel announces to young Mary: "You will conceive and give birth to a son, and you are to call him Jesus."[21]

Her reaction?

"'How will this be,' Mary asked the angel, 'since I am a virgin?'"[22]

This event is consistent with the messianic prophecy made by Isaiah many centuries before. "Look!" the ancient prophet wrote. "The virgin will conceive a child! She will give birth to a son and will call him Immanuel (which means 'God is with us')."[23]

Look! is a modern English translation of the Hebrew word *hinneh* (הנה) often translated as *behold*. "It is used to arrest attention; to indicate the importance of what was about to be said," explains Albert Barnes, a nineteenth-century American theologian. "It is used in lively descriptions, and animated discourse; when anything unusual was said, or occurred; or anything which especially demanded attention."[24]

The virgin is an English translation of the Hebrew phrase *ha'almah* (העלמה). *Almah* alone, says Barnes, "properly means a girl, maiden, virgin, a young woman who is unmarried, and who is of marriageable age."[25] But *ha'almah* means *the* maiden, which gives the young woman special significance.

There's an ongoing debate about whether Mary was indeed a virgin, in part because *'almah* can also be translated as *young woman* or *maiden.* But given the strict morality of Mary's day and culture, scholars point out, a young woman of marriageable age was almost certainly a virgin.

A virgin birth might sound far-fetched to many people; but it didn't to me.

First of all, it's pretty clear that the God of the Bible, if you believe he exists, has the power to do whatever he wishes. If he spoke an entire universe into existence, certainly he can speak life into a virgin's womb.

Second, I was well aware that many creatures in nature have the remarkable ability to reproduce asexually. Virgin births happen among reticulated pythons, bonnethead sharks, honeybees, brine shrimp, Komodo dragons, and even domesticated turkeys.[26]

The process is called *parthenogenesis.* It's commonplace, but we still don't fully understand it.

"It's amazing that we do all of this work on reproductive biology," remarks University of Tulsa geneticist Warren Booth, "and we're still learning something new about the reproductive modes about the animals around us."[27]

Natural virgin births have never been seen among *mammals*—certainly not humans. A "clinically normal, human parthenote (i.e., a healthy individual entirely derived from a single parthenogenetic-activated oocyte), has never been reported in the scientific literature."[28] However, in 2004, Japanese scientists at Tokyo University made history by genetically engineering the first virgin birth of a mouse. They named the miraculous little mammal Kaguya.[29]

THE LIFE OF JESUS

Believe whatever you like about Jesus, but he actually existed. That's what I soon discovered while investigating the Christian worldview.

In fact, there's more historical evidence for the life of Jesus than for many other towering figures of antiquity. This includes Socrates, Tiberias Caesar, and Alexander the Great, whom no reasonable person seriously doubts existed.

In addition to the twenty-seven letters bundled into the New Testament, there are eighteen or so historically authenticated, extrabiblical, non-Christian sources that together affirm the existence of an extraordinary person named Jesus—who, during the first century of the common era, disturbed the civil and religious peace of his day, was crucified for it, and whose followers claimed came back to life, spawning an even bigger, global, revolution that continues to this day.[30]

One of these independent sources is *The Antiquities of the Jews*, a twenty-volume Jewish history authored by Flavius Josephus, a Jewish aristocrat and historian born shortly after the time of Jesus' crucifixion. In the words of Bart Ehrman, a New Testament scholar at the University of North Carolina, "Flavius Josephus is far and away our best source of information about first-century Palestine."[31] Ehrman also argues that "there is no serious doubt for virtually any real scholar of antiquity (whether biblical scholar, classicist, historian) that Jesus of Nazareth really did live."[32]

Many (though not all) scholars suspect that some of Josephus's original references to Jesus were doctored by later Christians. The following references, however, are widely considered authentic:

> Now there was about this time Jesus, a wise man . . . ; a doer of wonderful works, a teacher of such men as receive the truth with pleasure. He drew over to him both many of the Jews and many of the Gentiles. . . . Pilate, at the suggestion of the principal men amongst us, had condemned him to the cross. . . . And the tribe of Christians, so named from him, are not extinct at this day.[33]

> Festus was now dead, and Albinus was but upon the road; so he assembled the sanhedrim [sic] of judges, and brought before them the brother of Jesus, who was called Christ, whose name was James, and some others. . . .[34]

Peter Shäfer is widely regarded as our era's preeminent scholar of ancient Judaism. In his meticulously researched book *Jesus in the Talmud*, Shäfer explains that leading rabbis of the late antiquity studiously avoided mentioning Jesus—and understandably so, given their antagonism toward the upstart religion he fathered. They saw Jesus as a phony, blasphemous troublemaker.

The few things they did say about Jesus smack of a disinformation campaign. As Shäfer explains, "The rabbis drafted . . . a powerful counternarrative [to the NT's narrative of Jesus] that was meant to shake the foundations of the Christian message: for, according to them, Jesus was not born from a virgin, as his followers claimed, but out of wedlock, the son of a whore and her lover; therefore, he could not be the Messiah of Davidic descent, let alone the Son of God."[35]

Yet the rabbis—and other ancient detractors who wrote about the Christian revolution—all agree on one thing: This rabble-rousing Jesus of Nazareth was as real as you and me. "Jewish rabbis who did not like Jesus or his followers accused him of being a magician and leading people astray," observes Purdue University's Lawrence Mykytiuk, "but they never said he didn't exist."[36]

Still, after all these centuries, some people clearly haven't gotten the memo.

In 2015, the Church of England polled 2,545 adults (18+) about their views of Jesus. They conducted a similar study in 2016 with 2,000 young people (ages 11–18). The study's findings were alarming: "40 percent of adults and 46 percent of young people in England either don't believe, or aren't sure, that Jesus was a real person who lived on earth."[37]

What an appalling level of ignorance about the single-most influential figure in human history.[38]

THE DEATH OF JESUS

I greatly enjoyed exploring Buddhism. But I remember feeling let down when I learned that the Buddha, Siddhartha Gautama, died of food poisoning—from eating either spoiled pork or toxic mushrooms.[39] It seemed like such a lowly, anticlimactic, pointless way for the Awakened One to exit the world.

According to the NT and other independent sources, Jesus' demise was also lowly: He was crucified alongside two common thieves. But the crucifixion wasn't pointless, far from it.[40]

According to the NT, Jesus' trial before Pilate was the mother-of-all court cases. For all the world to see and for all time, it represented the arrest, trial, conviction, and execution of our species, *Homo sapiens sapiens*, for its never-ending insurgence against God.

The NT explains: Only Jesus, God in the flesh, had the authority to deliver such cosmic justice; to end, once and for all, the interminable back-and-forth between our infractions and God's rebukes. Only Jesus, God in the flesh, the author of all humanity—past, present, and future—had the power to settle the score on behalf of all humanity, past, present, and future.

Above all, the NT says: Jesus' crucifixion was the *only way* God could demonstrate how much he loves us. "The greatest way to show love for friends is to die for them."[41]

According to the Old Testament, our species' criminal record stretches all the way back to the Garden of Eden. Adam and Eve defied God, and he delivered swift and just punishment:

> Then he said to the woman, "I will sharpen the pain of
> your pregnancy, and in pain you will give birth. And

> you will desire to control your husband, but he will rule over you."[42]
>
> And to the man he said, . . . "[T]he ground is cursed because of you. All your life you will struggle to scratch a living from it. It will grow thorns and thistles for you, though you will eat of its grains. By the sweat of your brow will you have food to eat until you return to the ground from which you were made."[43]

Yikes!

Afterwards, humanity chose to escalate the revolt; and God once again delivered swift and just punishment. According to the Old Testament, it happened during the lifetime of Noah, "a righteous man, the only blameless person living on earth at the time."[44]

Only Noah, his family, and representatives of Earth's flora and fauna were spared from the deluge of God's just sentence.

> When Noah was 600 years old, on the seventeenth day of the second month, all the underground waters erupted from the earth, and the rain fell in mighty torrents from the sky. The rain continued to fall for forty days and forty nights.[45]

According to the Old Testament, humanity continued to defy God, like incorrigible teenagers, and on each occasion received its just punishment. Even the Hebrews—God's chosen people—experienced a nonstop cycle of rebellion-punishment-contrition-forgiveness-rebellion-punishment-contrition . . .

From biology, I already knew that the human genome is riddled with 100,000 bits of DNA from the many viruses that have infected our species down through the ages. These ancient viruses are concentrated in the central regions of our chromosomes, called *centromeres,* which are devilishly complex. But now I couldn't help but think: If our chromosomes are infected with DNA from ancient viruses, it also appears—from my reading of the Old Testament—that we've inherited another ancient, infectious bug: the virus of Adam and Eve's rebelliousness.

This hoary pathogen is what Christians call *original sin*.

In light of all this, Laurel and I found the Old Testament to be rather depressing. It's all about our lawlessness and rebellion and God's anger and discipline. Loving our friends and hating our enemies. Seeking retribution: an eye for an eye. There's no happy ending. Worse yet, there's no clear possibility of there ever being a happy ending.

The Old Testament was depressing for another reason: It *rings true* in a trivial, logical, gloomy sort of way. It lacks hope that anything will ever change in the world—or inside of us.

Years earlier, I'd felt this same sinking feeling when exploring Judaism. I learned about *Tikkun olam* (תיקון עולם), which is Hebrew for repair, improve, or fix the world.[46] *Tikkun olam* is Judaism's noble-minded call to arms in the face of what is clearly a broken world.

Christianity, too, exhorts its adherents to do good works.[47] But unlike Christianity, Judaism relies *extra heavily* on us—on our strict adherence to the Mosaic laws and the global aggregation of *mitzvot*, good deeds—to make things right, here and now, and to determine our individual eternal fates. In short, it places an enormous trust in our rebellious species, in *human beings* who broke the world in the first place and keep breaking it with everlasting bad behavior.

Judaism is a worldview naturally steeped in the Old Testament, where God waits and waits and waits for us to shape up. But it hasn't ever worked—and it never will work because our rebellious nature is the problem, right down to our DNA. As long as you and I have *freedom of choice*, we will continue to make bad choices—raising hell, as it were, instead of making peace with heaven.

But in the Christian worldview, God already knows this—and he has a plan.

Before Creation, God knew there would only be one way out of humanity's endless, catastrophic cycle of rebellion and justice.

God knew there was only one way for *him* to remain just and for *us* to remain free.

God knew there was only one way to make peace, once and for all time, with his beloved, stiff-necked creation.

That *one way* is through the very public sacrifice of Jesus Christ, the incarnation of God himself.

"*I* am the way, the truth, and the life," Jesus says. "No one can come to the Father except through me."[48]

This is why early Christians called themselves "followers of the Way."[49] They recognized that Jesus represents the *only* way to solve the conundrum of justice versus freedom; the *only* way for an imperfect, chronically hateful, rebellious species to live in peace with a perfect, loving, and just God.

I confess it took me a long time to fully grasp this. My little, logic-centered worldview had a hard time wrapping its tiny arms around the paradoxical, translogical significance of the man-yet-God Jesus and his lowly-yet-heavenly sacrificial death.

But when I finally got it, it was a game changer. For me, it heralded the beginning of the end of my lifelong Atheism—and the end of the beginning of my intellectual and spiritual plunge into the deepest mysteries of the universe and life.

THE RESURRECTION OF JESUS

According to one of Jesus' most ardent followers, known to history as Paul the apostle, Christianity's legitimacy depends entirely on the answer to a single question about Jesus: *Did he or didn't he return from the dead?*

> If Christ has not been raised, then all our preaching is useless, and your faith is useless. And we apostles would all be lying about God—for we have said that God raised Christ from the grave.[50]

In a missive to the early Christian church in Corinth, Greece, Paul goes on to answer the critical question by avowing that Jesus did indeed rise from the dead. The letter has been reliably dated to about 55 AD—only twenty years or so after Jesus' resurrection. So it's like someone today writing about 9/11.

Paul reports:

> He [the resurrected Jesus] was seen by Peter and then by the Twelve. After that, he was seen by more than 500 of his followers at one time, *most of whom are still alive*, though some have died. Then he was seen by James and later by all the apostles. Last of all, as though I had been born at the wrong time, I also saw him.[51]

This extraordinary claim is significant for many reasons. For one, skeptics at the time could easily verify (or discredit) Paul's assertion by running down and questioning the hundreds of other eyewitnesses Paul alleges are still alive.

For another, Paul himself is a particularly credible witness. Born Saul of Tarsus, a devout, rabbinically trained student of Judaism, he became the most vicious persecutor of Christians in his day. He mocked Christians for believing in Jesus and hunted them down like animals all over the Mediterranean region.

The New Testament describes Saul's brutal behavior during the years immediately following Jesus' death and resurrection:

> A great wave of persecution began that day [with the killing of Stephen, a well-known follower of Jesus] . . . and all the believers except the apostles were scattered through the regions of Judea and Samaria. . . . Saul was going everywhere to destroy the church. He went from house to house, dragging out both men and women to throw them into prison."[52]

Historians agree that Saul of Tarsus actually existed and that his story is accurate. They also agree that something out of the ordinary happened to him a few years after Jesus' resurrection, around AD 37.[53] Something abrupt and surreal that instantly spun his life 180 degrees.

In the blink of an eye, Saul went from being a ruthless persecutor of Christians to *one of them*! Moreover, he went on to become Christianity's most articulate and prolific spokesperson, writing many of the works that make up the New Testament.

In one letter, Saul/Paul candidly recalls his old self: "I was so zealous that I harshly persecuted the church."[54] In another letter, he admits, "I violently persecuted God's church. I did my best to destroy it."[55]

Paul's credible testimony notwithstanding, I understand why many people today still struggle with believing that Jesus really, truly rose from the dead: *It sounds utterly fantastical*. Still, as a scientist, I'm used to taking seriously far-out sounding claims. You might say it's the bread and butter of modern science.

Quasars, for example. They're one of the great mysteries of the far-flung cosmos.

We believe them to be unusually bright galaxies centered on super-massive, killer black holes. But they're extremely far away, so our telescopes can't actually see them; so everything we claim to know about them is based on extravagant inferences made solely from the light they give off, which isn't much to go on.

Not surprising, then, our beliefs about quasars are riddled with uncertainties. Case in point: The most distant quasar ever discovered, ULAS J1342+0928, is supposedly eight hundred million times more massive than the sun and located 77,000,000,000,000,000,000,000 miles away, at the very edge of the reachable universe.[56] But there's a major problem with that widespread belief: The quasar's allegedly enormous size doesn't square with our standard model of cosmology—our best theory about the universe.

All that said, no serious-minded astronomer doubts that quasars exist or that, far-fetched as it sounds, quasars are what we say they are: unimaginably bright galaxies being systematically gobbled up by massive black holes.

As I investigated the credibility of what the Bible says about Jesus, I quickly realized there's far more hard evidence for his resurrection than there is for quasars generally and ULAS J1342+0928 in particular.

E. P. Sanders, Duke University's eminent religious scholar, now-retired, avows in his landmark book *The Historical Figure of Jesus*:

> There are no substantial doubts about the general course of Jesus' life: when and where he lived, approximately when and where he died, and the sort of thing that he did during his public activity. . . .
>
> We may add here a short list of equally secure facts about the aftermath of Jesus' life: "his disciples at first fled; they saw him (in what sense is not certain) after his death." . . .
>
> We know that after his death his followers experienced what they described as the "resurrection": the appearance of a living but transformed person who had actually died. They believed this, they lived it, and they died for it.[57]

Over the centuries, skeptics have offered up alternative, naturalistic explanations for Jesus' documented reappearances after his crucifixion. But Gary Habermas, a leading academic authority on Jesus' resurrection, has rebutted every one of them using what he calls a "minimal facts approach," which relies only on facts that meet two criteria: "They are well evidenced and nearly every scholar [who studies the subject] accepts them." In other words, says Habermas, "We present our case using the 'lowest common denominator' of agreed-upon facts."[58]

When you "use the New Testament the way skeptics use the New Testament," Habermas says—that is, "if you use [only the minimal] data scholars allow—[there's no doubt] Jesus has been raised from the dead."[59]

As I considered this argument, I realized something else. By the time you get to this point in Jesus' biography, your mind is probably firmly made up about him.

If you believe it's possible that God caused Mary to become pregnant while still a virgin, then honestly, it's no big deal to believe that God could resuscitate a dead Jesus. On the other hand, if you disbelieve the virgin birth, you're not likely to believe in Christ's resurrection.

Ultimately, with apologies to Paul, I judged the Christian worldview not solely on the question about, and the evidence for, the resurrection of Jesus but on the best available evidence concerning his *entire* story—from the alleged prophecies of his birth to his alleged return from the dead.

It took me years to do this because I—Michael Jillion—needed to settle a jillion questions before coming to a decision. But one day it finally became clear to me what that conclusion had to be.

It wasn't an emotional experience for me. Rather, it was the culmination of an intellectual dawning, a gradual awakening, that had begun two decades earlier at Cornell University, when I—an unkempt, malnourished scientific monk—asked myself a simple but pointed question: *How did this amazing, mostly invisible universe of ours come to be so amazing and mostly invisible?*

The answer, I now concluded, had something to do with the magnificent and enigmatic stars, galaxies, and quasars I'd studied all my life. But it also had *everything* to do with the loving God who spoke them into being . . . and the resurrected Jesus, who brought this loving but remote God down to Earth, making it possible for me—for you, for anyone—to know him *personally*. To make heavenly peace with him, at last.

> For ever since the world was created, people have seen the earth and sky. Through everything God made, they can clearly see his invisible qualities—his eternal power and divine nature. So they have no excuse for not knowing God.[60]

BELIEF AND TRUST

What does the NT mean when it says that Jesus beckons you to *believe in him*? And why is your response to this invitation so central to the Christian worldview?

How can *one decision*—a decision about someone who popped in and out of our world two thousand years ago—possibly spell the difference between eternal life and eternal death for you? And what exactly do those two outcomes mean, anyway: *eternal life* and *eternal death*?

These are just a few of the questions that roiled my thoughts as I strove to understand Jesus and Christianity. Even in grad school and while teaching at Harvard, I was still that kid who drove his teachers crazy with his unrelenting curiosity and endless queries.

Gradually, my questions about the prophecies concerning

Jesus—and his birth, life, death, and resurrection—herded me toward revelations quite unlike those I'd gotten from any other religion—yet very much like those I'd gotten from the sciences I loved so much. These revelations radically and forever expanded my worldview and transformed my life.

One such revelation taught me the quantum difference between *believing* in something (with mere IQ-powered faith or mere SQ-powered faith alone) and *trusting* in something with full-bodied IQ-and-SQ-powered faith, the mightiest faith of all. And the difference between believing something exists and *surrendering* yourself entirely to it, body, mind, *and* spirit.

The New Testament explains the difference by making an astonishing claim:

> This is how God loved the world: He gave his one and only Son, so that everyone who believes in him [i.e., trusts him *completely*] will not perish but have eternal life.[61]

This verse isn't speaking about some trivial kind of *belief*. Rather, it envisions a deep, meaningful *trust*—that is, an "assured reliance on the character, ability, strength, or truth of someone or something."[62]

Here's another way to understand it.

At my high school in Southern California, students had to pass a swimming test in order to graduate. This was a huge problem for me because I had a deathly fear of water.

I'll never forget when the swim coach made me get into the deep end of the pool. Saucer-eyed and shivering all over, I held on to the edge of the pool with a death grip.

"Son, there's nothing to be afraid of," the coach told me. "Just relax and let go!"

But every time I tried, I instantly started sinking like a stone.

Ironically, as an uber geek, I knew all about Archimedes'

principle, the scientific law of buoyancy. I could recite its main equation off the top of my head and calculate the buoyant force, F_b, on any physical object imaginable—including my own skinny, frightened self:

$$F_b = \rho \times g \times V$$

I *believed* in this equation. I *believed* in Archimedes' principle. I *believed* in the science behind it.

But they were all beliefs powered merely by my geeky IQ-based faith.

I didn't *trust* them with my life.

I didn't trust the phenomenon of buoyancy with my whole IQ-SQ self.

That's why I kept sinking. *Literally.*

For reasons I won't get into, I managed to graduate from high school without ever learning to swim—much to my everlasting regret, I might add. Over the years, I've missed out on the joy of participating in water-related recreation with my beloved son, who swims like a fish.

I did, however, learn how to float—even on my back, which for me is a big deal. Now, every time I jump into a pool—no more than five feet deep!—I consciously relax my muscles and completely surrender myself to the water. I've learned it's the *only way* I can experience the joy of buoyancy.

Years after the Christian worldview opened my eyes to the vast difference between *belief* and *trust,* I still struggled with putting that awareness into practice. You might say I relived my high-school predicament—except this time, it was with Jesus, not water.

I now realize I was relying too much on my IQ and not enough on my SQ to fully trust Jesus. You might say my book smarts, of which I was very proud, were getting in the way.

Here's how the NT explains the problem of IQ lording it over SQ:

> The person without the Spirit does not accept the things that come from the Spirit of God but considers them foolishness, and cannot understand them because they are discerned only through the Spirit.[63]

One day, however, when I was feeling beaten down by a crisis I'll describe to you shortly, it happened: After decades of being an Atheist-scientist, of assiduously exploring the world's great religions, I relaxed . . . let go . . . and completely surrendered myself to the buoyancy of Jesus' promise:

> Come to me, all of you who are weary and carry heavy burdens, and I will give you rest. . . . Let me teach you, because I am humble and gentle at heart, and you will find rest for your souls.[64]

At that fateful moment, instead of continuing to sink deeper and deeper into despair, I suddenly felt myself floating. My spirits were instantaneously lifted, buoyed by an ineluctable joy that made me feel *alive* as never before.

It marked the first time in my years-long investigation of the NT that I truly comprehended one of its most puzzling, profound, translogical truths. As Jesus told his disciples: "Whoever would save his life will lose it, but whoever loses his life for my sake will find it."[65]

For the first time, I grasped the cosmic gravity and truth of the audacious words that Jesus spoke to a grieving woman whose brother had just died: "I am the one who brings people back to life, and I am life itself. Those who believe in me will live even if they die."[66]

REBIRTH

Central to the Christian worldview is a pair of brazen ideas:

- Jesus has the power to restore the physical, mental, and spiritual life and well-being of any person.
- The key to unleashing this power is for the person to *trust* Jesus; to surrender their entire IQ-SQ being to him—body, mind, and spirit.

The NT contains many reports of individuals, named and unnamed, who experienced miraculous recoveries and transformations. One of them was Paul (formerly Saul of Tarsus), whose deeply changed life became a living testimony to the overwhelming power of God. In fact, Paul told anyone in the ancient Greco-Roman world who would listen: *Jesus has the power to radically transform a person's life.*

In one of his letters, Paul makes a point of explaining that his conversion wasn't powered by an IQ-related change of mind or heart. It was powered by an SQ-related source that can't be seen, proven, or even imagined:

> I want you to understand that the gospel message I preach is not based on mere human reasoning. I received my message from no human source, and no one taught me. Instead, I received it by direct revelation from Jesus Christ. . . .
>
> God chose me and called me by his marvelous grace. Then it pleased him to reveal his Son to me so that I would proclaim the Good News about Jesus to the Gentiles."[67]

Paul was now a new person, completely reborn. And if it could happen to a wretch like him, Paul exclaimed, it could happen to *anyone*, including you and me: "Anyone who belongs to Christ has become a new person. The old life is gone; a new life has begun!"[68]

Whatever you believe about Christianity, it's nearly impossible—certainly unreasonable—to shrug off Paul's claims. No one willingly puts his life on the line for something he doesn't sincerely believe.

Paul, the erstwhile persecutor, suddenly found *himself* being persecuted for being a Christian. He was abused, imprisoned more than once, and ultimately beheaded in Rome, reportedly on the orders of Emperor Nero.[69]

In one of his letters to the church in Corinth, Paul explained his willingness to endure such pain and persecution:

> Five times I received from the Jews the forty lashes minus one. Three times I was beaten with rods, once I was pelted with stones, three times I was shipwrecked, I spent a night and a day in the open sea, I have been constantly on the move. I have been in danger from rivers, in danger from bandits, in danger from my fellow Jews, in danger from Gentiles; in danger in the city, in danger in the country, in danger at sea; and in danger from false believers. I have labored and toiled and have often gone without sleep; I have known hunger and thirst and have often gone without food; I have been cold and naked. Besides everything else, I face daily the pressure of my concern for all the churches. Who is weak, and I do not feel weak? Who is led into sin, and I do not inwardly burn?
>
> If I must boast, I will boast of the things that show my weakness. . . .
>
> In order to keep me from becoming conceited, I was given a thorn in my flesh, a messenger of Satan, to torment me. Three times I pleaded with the Lord to take it away from me. But he said to me, "My grace is sufficient for you, for my

> power is made perfect in weakness." Therefore I will boast all the more gladly about my weaknesses, so that Christ's power may rest on me. That is why, for Christ's sake, I delight in weaknesses, in insults, in hardships, in persecutions, in difficulties. For when I am weak, then I am strong.[70]

During the many years I investigated the Christian worldview, I never experienced a sudden epiphany quite like Paul's on the road to Damascus.[71] My conversion was more like a slow awakening; a systematic march toward well-thought-through conclusions that were consistent with the best available evidence.

What else would you expect from a lifelong atheistic, scientific monk?

I did have one experience, however, that came close to becoming an abrupt moment of surrender. It is forever seared in my memory.

I'm reluctant to recount it for you because, as a hard-nosed scientist, I've always listened skeptically to other Christians' conversion stories. It's not that I think they're fibbing or stretching the truth; it's just that in many cases there's no way to test or refute what they say. Their highly personal narratives are not legitimate scientific hypotheses.

But I've decided to tell you anyway, partly because it's the honest thing to do. And partly because the event radically and forever changed my life, my worldview, my *everything*.

It began in 1991, when I married Laurel, the young woman who had upended my monastic existence at Cornell with a simple Valentine's Day card. Being Latino and having a Cuban/Mexican-American mother who was the eldest of eight siblings (four boys and four girls), I wanted to have lots of kids. Happily, so did Laurel.

However, it didn't take long for us to discover we couldn't have children. A top fertility doctor in Boston informed us that we fell into a category called "unexplained infertility," meaning that we both checked out fine biologically.

The mystery of our plight gave us hope. We turned to artificial insemination (AI) at one of the best fertility clinics in the world; but it failed. Undaunted, we tried again, but it failed. We tried over and over again, but the outcome was always the same: *nothing*.

All this was happening while Laurel and I were seriously studying Christianity. We were trying extra hard to be faithful to the religion.

I'll never forget our first attempt at AI. A nurse and I stood next to Laurel, who was lying on an examining table inside a small room. Just before the procedure, the nurse opened a drawer, pulled out a wooden idol, and invited us to kiss it. "For good luck," she said, smiling.

Laurel and I both recoiled and told the woman to put the idol away. It gave me the creeps—instantly darkening my spirit—so I uttered a short prayer for a successful pregnancy. That prayer was not answered—nor were many, many, many more prayers like it.

After several years of unremitting failure, Laurel and I were worn out physically, emotionally, mentally, and spiritually. I won't speak for Laurel—she's a brave, wise woman—but I began thinking the Christian worldview was not legit; that trusting Jesus had no real positive effect.

The sense of disappointment was all too familiar; I had experienced it with every other religion I'd ever explored. Their claims—including Transcendental Meditation's promise that I could levitate, if only I *believed*—had never worked for me.

My personal crisis became so grave that I couldn't sleep at night. I've always been a thin person, but I grew even skinnier and increasingly depressed.

We were living at the time on a lovely country property in rural central Massachusetts, about an hour's commute to Harvard. Our second-floor master bedroom was large enough to include my office, and it had a stunning view of our property's gently rolling meadows. We called it Tranquility Fields, after the Apollo moon landing site, Tranquility Base, but I felt anything but tranquil.

Late one afternoon in the fall of 2003, I sat at my desk looking gaunt and feeling heavy laden. Through the large, mullioned picture window, I stared out at the fields of withering wildflowers. They looked the way I felt inside: *lifeless.*

Impulsively—desperately, actually—I rang up a pastor friend of mine and spilled my guts to him for nearly an hour. After listening quietly, he spoke words I'll never forget: "Brother Michael, you need to let this all go, it's killing you. You gotta surrender it—the whole thing—to Jesus."[72]

I heard his words, but I didn't really hear what he was saying to me.

Thanking him, I hung up the phone and sat there once again, stewing in my disappointment, anger, and unbelief. Minutes later, totally spent, I shuffled toward my bed and collapsed onto it like a sack of rocks, fully clothed.

That's when it happened.

Even now, I find it hard to put it into words—like Joe the split-brain patient who couldn't verbally explain what he was looking at, even though it was real and right in front of him.

Years later, I came to recognize this ineffability as a telltale sign of profound, translogical, SQ-based truths, realities, and experiences. They're impossible to describe with words, logic, or even a lively imagination. By contrast, trivial, logical, IQ-based truths, realities, and experiences are generally quite easily put into words.

"When we tell you these things," Paul said, explaining the Christian worldview, "we do not use words that come from human wisdom. Instead, we speak words given to us by the Spirit, using the Spirit's words to explain spiritual truths."[73]

The Austrian-British philosopher and logician Ludwig Wittgenstein sought to understand the fundamental difference between the *profound*, *translogical*, *inexpressible*, and the *trivial*, *logical*, *expressible*. His conclusion? If I seek to understand the universe with my IQ alone, then "*the limits of my language* mean the limits of my world."[74]

For my entire life, I had been that person. With the notable exception of quantum physics, my tiny worldview was chock-full of truths, realities, and experiences that were strictly logical and therefore relatively trivial. They were truths, realities, and experiences that for the most part were expressible in words.

But on that late fall afternoon in New England, my worldview exploded. And it has never shrunk back to what it once was.

As I fell into bed, I suddenly beheld the image of a large, translucent, humanlike figure. Its giant arms caught and embraced me. I *felt* the embrace. In other words, what I experienced was both visual and tactile.

A split-second later, as my head was about to hit the pillow—this is the part of the experience that affected me the most—I came cheek-to-cheek with the humanlike figure. *I felt it.* It was warm and soft.

The unexpected, electrifying sensation; the abrupt and seismic spiritual awareness—words really do fail me here. Whatever it was that happened to me at that moment is indescribable.

Perhaps the best I can do is to quote Rabbi Abraham Joshua Heschel, the towering twentieth-century Jewish scholar: "Obviously, we can never sneer at the stars, mock the dawn or scoff at the totality of being. Sublime grandeur evokes unhesitating, unflinching awe. . . . Standing between earth and sky, we are silenced by the sight."[75]

It wasn't just a spiritual experience. It was a physical one as well. Within moments of my head hitting the pillow, it was *light's out*. For the first time in months, I slept like a baby.

When I awoke in the morning, I was tranquil and rested, as if I'd vacationed for two weeks in the Bahamas. When I sat up in bed and looked around my room, it felt as if nothing had changed—Laurel and I were still childless. And yet *everything* had changed.

It was a translogical feeling as paradoxical as the idea of a quantum vacuum that is at once empty and full. A reality that defied logic as much as it did language.

Did my mysterious encounter *prove* that Jesus is real? *No.* Could I think of perfectly natural explanations for it? *Yes.*

But what I couldn't deny—not then, not now—is that the experience was wholly consistent with the best available biblical and extra-biblical evidence and the mainstream Christian worldview.

And it didn't stop there.

Almost immediately, the thoughts, people, and circumstances in my life underwent a sea change. I wish I had the space to explain it all, but let me give you one example.

One Sunday morning soon after my inexplicable encounter, Laurel and I were driving home from church. Without warning and without either of us speaking a word, we turned to one another with excitement.

"What?" I said.

"I think we're supposed to adopt!"

I gasped. "That's what I just realized too!"

We marveled at how we'd both suddenly felt wordlessly prompted by something or someone inside the car—*at precisely the same moment.*

Years later, when I wrote *Amazing Truths: How Science and the Bible Agree*, I included a chapter about how both the Bible and science believe in the possibility of instantaneous communication between widely separated parties. In physics, it's called *quantum entanglement.* In Christianity, it's known as the Holy Spirit.

Call it what you will, but on that Sunday morning, it's quite possible that Laurel and I experienced it firsthand.

Up until then, I'm embarrassed to admit, we'd never seriously considered adopting a child. Partly it was ignorance; partly fear.

Lots of people, we discovered, feel the same way. When I mentioned adoption to a friend, he quickly pushed back and said, "What if you get a bad seed?"

But on that auspicious Sunday, there was suddenly no doubt in our minds. *None.* Every bit of fear and ignorance was totally gone, like a morning fog burned away by the sun.

Laurel and I were so excited that we rushed straight home, booted up the computer, and spent the rest of the afternoon and evening studying the adoption process required by the Commonwealth of Massachusetts. We didn't even stop to eat, and that night we hardly slept.

At eight o'clock sharp the next morning, I called the Department of Children and Families and started the ball rolling. Soon Laurel and I were taking night classes on parenting; being visited by a social worker who measured the square footage and safety of our two-story home (we had enough room for a passel of kids!); being interviewed individually and together; being vetted by the police; attending adoption parties on weekends; and meeting kids of all stripes who were waiting for a forever home.

At times we wanted to adopt them all!

I won't kid you: Sometimes it got scary. But not once—even at the scariest moments—did we ever waver from the conviction that this was meant to be; that we were doing exactly what God wanted us to do.

That faith emboldened us. The NT puts it this way: "There is no fear in love, but perfect love casts out fear."[76]

Despite everything, that wild ride was one of the happiest experiences we've ever had. For the first time in what seemed like forever, Laurel and I felt alive—truly *alive*!

Even as I write these words, I'm tearing up—and smiling ear to ear—as I'm reminded of the sheer vitality and joy of that experience. And also of those audacious words that Jesus spoke to the bereft woman who'd lost her beloved brother two millennia ago: "I am the one who brings people back to life, and I am life itself. Those who believe in me will live even if they die."[77]

After about a year of waiting to be matched with a child, we received a call one Friday evening around five. I'll never forget it. I picked up the phone, and a young social worker said to me: "I think I have the perfect little four-year-old boy for you."

Those were her exact words.

And she was right.

We met the little guy, instantly fell in love with him, and he with us. When I first heard his sweet voice—even before I laid eyes on him—I turned to Laurel and said, "That's my son."

Once again, those were the *exact* words. I remember it like it was yesterday.

Our precious son is in his very early twenties now and is the love of our lives. It's a love that transcends Darwinian love. A love that transcends mere DNA.

It's a divine love that Laurel and I had never experienced before meeting him. A love described in the New Testament this way: "We know how much God loves us, and we have put our trust in his love. . . . And as we live in God, our love grows more perfect."[78]

In speaking to others about our experience, I'll often say that Laurel, our son, and I are a family made in heaven. And I don't mean it figuratively; I believe it's literally true.

Every time I look into my son's eyes, I see God flashing me a smile that says, "See, I do keep my promises—if only you'll get out of the way, surrender your will, and give me the chance to do so."

I tell my boy as often as I can: "I love you, son." And I mean it. He's the most wonderful child any parent could ever hope to have.

It's also my way of saying, "I love you, God. I love you, Jesus. *Thank you.* A million times a million times, *thank you.*"

FREEDOM

Atheists are wont to boast about being *free thinkers*. Of not buying into what they regard as feeble-minded, Bronze Age religious fables and superstitions.

It raises a legitimate question: Are they right? Does religion—does Christianity, one of Atheism's favorite targets—enslave people with myths and mysteries? Does Atheism liberate people?

Setting aside the fact that Atheism itself is a religion—founded on axiomatic beliefs that are at odds with science and cannot be proven—I can say with confidence that devout Hindus, Buddhists, and Jews don't for a minute agree that Atheism is liberating.

Christians certainly don't. From a Christian perspective, Atheists are the most enslaved, least free-minded people of all.[79]

Let me explain by returning to the central thesis of this book. You and I, as *Homo sapiens sapiens*, have a cognitive superpower, an SQ, that senses, albeit imperfectly, the existence of an unseen, immaterial universe that includes the idea of God. SQ gives us incomplete glimpses of God and of countless other profound, translogical truths and realities hidden from our eyes, our IQ, and our imagination.

SQ gives even Atheists a sense of God's existence,[80] though most will strenuously deny it. Or they'll go out of their way to credit their SQ to anything but God—such as the *universe*, *Gaia*, or *The Force*.

At most, some Atheists openly confess to a *feeling* they can't quite put into words and that IQ alone cannot explain.

According to the Pew Research Center, about 18 percent of American Atheists believe in a "higher power." And 31 percent of American Atheists say they "often feel a deep sense of spiritual peace and well-being."[81]

Consider the ruminations of one Staks Rosch, a contributor to the *Huffington Post* who describes himself as "a vocal atheist, humanist, progressive, and Jedi." In a 2017 editorial, Rosch struggles with the concept of spirituality before finally concluding, "To put it simply, spirituality is the feeling of deep connection we have towards one another and with the universe in general."[82]

So for Rosch and other like-minded Atheists, your spirituality is nothing more profound than your relationship with other people and the material universe. That is, your spirituality boils down to nothing more than mere emotionality—including that universal sense of wonder we all feel whenever we look up at the stars at night.

In 2007, neuroscientist and popular Atheist Sam Harris caught flak from many attendees of the Atheist Alliance conference in Washington, DC, when he delivered a speech titled "The Problem with Atheism."

> One problem with atheism . . . is that it seems more or less synonymous with not being interested in what someone like the Buddha or Jesus may have actually experienced. In fact, many atheists reject such experiences out of hand, as either impossible, or if possible, not worth wanting. . . .
>
> As atheists, our neglect of this area of human experience puts us at a rhetorical disadvantage. Because millions of people have had these experiences, and many millions more have had glimmers of them, and we, as atheists, ignore such phenomena, almost in principle, because of their religious associations—and yet these experiences often constitute the most important and transformative moments in a person's life. Not recognizing that such experiences are possible or important can make us appear less wise even than our craziest religious opponents.[83]

Ouch.

In 2014, Harris wrote a book called *Waking Up: A Guide to Spirituality without Religion*, in which he doubles down on his critique of Atheism's small-mindedness.[84] Here's the publisher's summary of Harris's basic thesis:

> "*Waking Up* is for the twenty percent of Americans who follow no religion but who suspect that important truths can

> be found in the experiences of such figures as Jesus, Buddha, Lao Tzu, Rumi, and the other saints and sages of history. Throughout this book, Harris argues that there is more to understanding reality than science and secular culture generally allow."[85]

I now recognize in my former self a Harris-style Atheist. I was a young man who sensed there was more to reality than what I was learning in my basement lab and my physics, math, and astronomy classes; and who eventually set out to discover what that extra-scientific, extralogical universe was like.

After years of searching—after years of probing the world's religions, ending grudgingly with Christianity—I've now concluded that our seminal fate as *Homo sapiens sapiens* is to be at peace with God, our creator, in a state of immortality and complete freedom.

The Bible portrays this idyllic state quite accurately, I believe, in its description of the Garden of Eden. The place where the newly minted Adam and Eve—made in God's likeness to be *at one* with him—were ageless and allowed to be free spirits.

Why did God allow them to make autonomous choices? Because he loved them—his most exquisite creations—too much to clamp down on their behavior.

As the father of a Gen-Z son, I completely understand God's decision. If you truly love your grown children, you will voluntarily, if reluctantly and anxiously, limit your authority over them.

Such an act of love, of course, automatically invites trouble into paradise. It creates the possibility that your loved ones won't listen to your advice; that they'll make bad decisions—and, worse, have to suffer the consequences for them.

God easily could've kept such trouble out of the Garden of Eden by lording over our species; by forcing himself on us. But he, more than anyone, understands that *control* is inimical to *love*.

Love is why God planted the tree of the knowledge of good and evil—the tree of free choice—smack in the middle of the garden.[86] *Love* is why, like a good parent, he warned Adam and Eve about it: "If you eat its fruit, you are sure to die."[87]

Free thinking is why Adam and Eve disobeyed. *Free thinking* is why they bought into the lie that God was a dissembling tyrant.

You and I are no different. We'll gladly ditch our peace with God to indulge our favorite pleasures—even if afterward we feel remorseful.

You and I believe the lie that in order to be ourselves—to be truly free thinkers and free doers—we must *escape* from God's iron-fisted, party-pooping clutches—or, as the Atheists do, escape from the very *idea* of God.

But Christianity believes just the opposite—that we experience *real* freedom, *perfect* freedom, only when we *draw near* to God.

It's one of Christianity's many puzzling, translogical truths. And I saw that it has an intriguing, eye-opening counterpart in a branch of nuclear physics called *quantum chromodynamics* (QCD).

We scientists believe the universe is pushed and pulled by four different forces. The strongest one—aptly named the *strong force*—holds together atomic nuclei the way corn syrup holds together popcorn balls.

The strong force is at the very core of our physical being. It's what holds us together. Without it—*poof!*—we'd disintegrate and blow away like dust in the wind.

And that's not all.

The strong force is a profound mystery. It originates deep within atomic nuclei, out of sight, in what we call *quarks*, the invisible and enigmatic components of protons and neutrons.

No one's ever seen a quark. And if QCD is correct, no one ever will.

Using the most powerful particle accelerators in the world, we've bombarded the heck out of atomic nuclei in hopes of busting loose a quark—but with no luck.

Now, you might think that if we ever did succeed in "liberating" a quark, it would be truly *free*—just as Atheists think that liberating themselves from any notion of God makes them truly free thinkers.

But just the opposite is true.

According to QCD, quarks stay within atomic nuclei because that's where they're *most* free. Let me put it another way. A quark is perfectly free only when there is *zero* distance between it and the source of the strong force. We call this extraordinary phenomenon *asymptotic freedom*.[88]

The strong force, in other words, *eases* its grip with decreasing distance—which is the exact opposite of how the other forces operate. Gravity and electromagnetism *strengthen* their grip with decreasing distance.

So a quark is freest when it sticks close to the nucleus and least free when it's farthest from the nucleus. In other words, the presence of the strong force is liberating, not enslaving.

According to Christianity, the identical thing is true about God. His presence is liberating, not enslaving.

How exactly is it liberating?

Christians believe that translogical realities such as God, Jesus, the Holy Spirit, angels, demons, satan, heaven, and hell truly exist. Atheists do not.

Atheists liken all of the above to obvious fictions such as the Tooth Fairy, Santa Claus, unicorns, and hobbits. And they do so with a sense of enormous pride, claiming to believe only in things they can see, prove, imagine, and support with solid evidence.

But as we've established throughout this book, we cannot see 95 percent of the observable universe nor 100 percent of the unobservable universe beyond that; truth is bigger than proof; logic can only prove trivial truths, not translogical truths; evidence can always be interpreted in more than one way, and it can never *prove* a complex hypothesis or religion, including Atheism.

The best we can ever hope to say about any belief is that it is *consistent with the best available evidence*, be it solid or ambiguous. Limiting ourselves to beliefs that can be seen, proven, and imagined is nothing to boast about.

In fact, having such a small worldview, as Atheism advocates, is symptomatic of a cognitive weakness, not a strength. It shows that we've closed our minds to the very truths that matter most: the profound, translogical truths that involve things that *cannot* be seen, proven, or even imagined—truths upon which both *science* and *Christianity* are solidly built.

Having such a small worldview, as Atheism advocates, shows we've closed our minds to invisible, logic-shredding, unimaginable possibilities such as the quantum vacuum and biblical creation, light photons and God, quarks and Jesus. Possibilities that are beyond words—which, as a scientist *and* a Christian, I take very seriously.

Atheism, then, not only celebrates small-mindedness; it is patently unscientific and un-Christian. It represents not free thinking but narrow, constricted thinking—thinking that is forever enslaved by the five human senses and simple logic.

Paul says essentially the same thing using different language:

> The natural man does not accept the things that come from the Spirit of God. For they are foolishness to him, and he cannot understand them, because they are spiritually discerned.[89]

In this same letter to the church in Corinth, Paul draws a distinction between the *spirit* (lowercase), which is the one we're naturally born with and that powers our SQ . . . and the *Spirit* (uppercase), which is the Spirit of God, the Holy Spirit that enlightens us when we're *reborn* spiritually.

Put another way, our standard-issue *human* spirit is profoundly augmented when we use our mighty freedom of choice to accept

God's full-blown Spirit—the Spirit he offered to Adam and Eve and that he offers to you and me. The Spirit with the power to explode worldviews and open small minds to the deepest translogical truths and realities of the universe—like some big bang of consciousness.

> No one can know a person's thoughts except that person's own spirit, and no one can know God's thoughts except God's own Spirit. And we [who have accepted God's offer] have received God's Spirit (not the world's spirit), so we can know the wonderful things God has freely given us.
>
> When we tell you these things, we do not use words that come from human wisdom. Instead, we speak words given to us by the Spirit, using the Spirit's words to explain spiritual truths.[90]

We receive God's Holy Spirit, the NT explains, the moment we acknowledge our imperfections; accept the forgiveness, love, and freedom that Jesus offers; *and* choose to trust him with our double-barreled IQ-and-SQ-based faith. Trust him with everything we have.

> "You must love the Lord your God with all your heart, all your soul, and all your mind." This is the first and greatest commandment.[91]

This is the critical moment of surrender, the moment when we *voluntarily* return to God the control he voluntarily forfeited, out of love for us, in the Garden of Eden.

It's the moment, above all, when the Holy Spirit makes it possible for you and me to have a *personal* relationship with God. Yes, I know. As audacities go, this is a big one.

When exploring Judaism, I learned that Jews consider it a sacrilege to believe that a mere, imperfect mortal can have a personal relationship with the great I AM. And I agree with them, in a way. How can you possibly cozy up to the creator of the universe when he's nothing more than a spiritual being *out there* somewhere?

Similarly, New Agers worship the universe "out there" and often say things such as: "If you really want something, just ask the universe for it." I had a producer at *Good Morning America* who constantly said that, and it always left me feeling cold and bewildered.

As a scientist, I've spent my entire life studying the universe and consequently feel close to it. But even so, I'd be zany to claim I could have a meaningful, personal relationship with galaxies, planets, and asteroids any more than I could with a supposed god who cares nothing about the daily affairs of his beloved creation.

In the Garden of Eden, Adam and Eve certainly had a close, personal relationship with God. In fact, because of their extreme closeness to God, you might say they were like a pair of quarks experiencing asymptotic freedom—*perfect, infinite freedom.*

When they defied God's single warning—"you must not eat from the tree of the knowledge of good and evil, for when you eat from it you will certainly die"[92]—they instantly became like quarks pulling away from the nucleus. They lost their perfect freedom.

People in Jesus' day could also legitimately claim to have a close, personal relationship with God—assuming Jesus was telling the truth when he addressed an audience of devout Jews and dropped this bombshell on them: "The Father and I are one."[93]

Immediately upon hearing this, the NT reports, "the people picked up stones to kill him."[94]

But Jesus, unfazed, tells them:

> "At my Father's direction I have done many good works. For which one are you going to stone me?" They replied, "We're stoning you not for any good work, but for blasphemy! You, a mere man, claim to be God."[95]

What about us today?

Jesus no longer walks the earth, so how can anyone living today, in the twenty-first century, possibly have a personal relationship with God?

Jesus explains:

> "All this I have spoken while still with you. But the Advocate, the Holy Spirit, whom the Father will send in my name, will teach you all things and will remind you of everything I have said to you."[96]

To my scientifically inclined mind, the Holy Spirit is like the *strong force*: It's real and extremely intimate. The Holy Spirit is hidden deep, deep within us; within the very quarks of our being. It doesn't get more intimate than that.

> The Kingdom of God does not come in such a way as to be seen. No one will say, "Look, here it is!" or, "There it is!"; because the Kingdom of God is within you.[97]

The Holy Spirit makes it possible for *you* to have an infinitely close personal relationship with God. A relationship in which there is *zero distance* between you and him. A relationship where you can experience what Adam and Eve once had: *perfect, asymptotic freedom.*

The Holy Spirit is not just SQ, a cognitive superpower. The Holy Spirit is a wireless, quantum-entanglement-like, broadband communications pipeline directly to the one who designed you. The one who designed a universe more than 546,712,159,706,000,000,000,000 miles across and decorated it with planets, stars, and galaxies to make you look up and wonder how such magnificence could possibly happen on its own?

The Holy Spirit—which the NT also calls the Helper or

Advocate—"will teach you everything and will remind you of everything" Jesus has said.[98] The Spirit will help you get to know the God who loves you so much that he voluntarily gives you the freedom to *invite* or *disinvite* him into your life. He offers you the opportunity to *choose* to follow his lead . . . or Adam's and Eve's.

According to the Christian worldview, following the wordless but perceptible promptings of the Holy Spirit—*trusting* Jesus' promises—keeps you close to the nucleus of God's presence. There, within the sound of your Creator's still small voice, you are at peace with him.

In the nucleus, you are *not* God's slave, as Atheism claims; rather, you exist in a state of asymptotic freedom, a state in which your IQ and SQ collaborate synergistically, divinely. This, in turn, empowers you to behold your life in all its logical *and* translogical magnificence and mystery.

Welcoming and entering that state of extreme closeness with God is the *only way* you'll ever be able to escape from yourself—from your rebelliousness, your bad habits, your self-deifying conceits.

As Jesus said, it's the *only way* "you will know the truth, and the truth will set you free."[99]

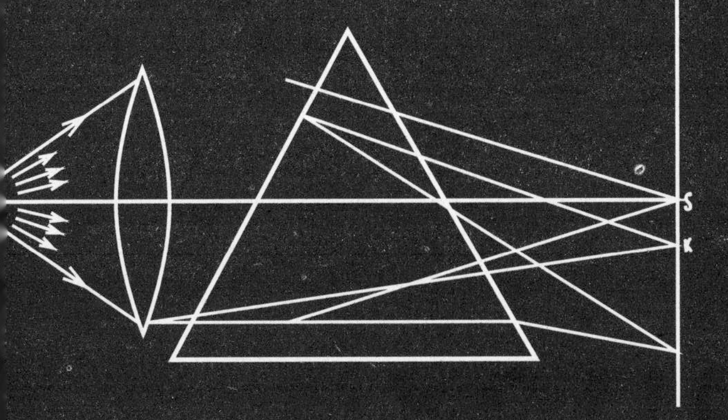

YOUR DESTINY

"Each of us is born seeing the world in a different way, and each moment we live shapes our eyes and hearts differently."

GREG KEYES

12

HAVING FAITH IN YOUR WORLDVIEW

Everyone in the streets said,
"How incomparable are the Emperor's new clothes!
What a train he has to his mantle! How it fits him! . . ."
"But he has nothing on!" a little child cried out at last.

HANS CHRISTIAN ANDERSEN

A YouTube video with millions of views shows a reporter named Joseph Backholm interviewing students at the University of Washington. These presumably intelligent young people are receiving a high-priced education at one of our nation's top public universities. They could easily be you, your sons and daughters, or your grandkids.

Backholm is an average-looking, 5'9", blue-eyed, white male in his late thirties. At one point he asks a female student, "If I told you that I was a woman, what would your response be?"

Without blinking an eye, the student replies cheerily, "Good for you, okay, like—yeah!"

Another student replies very politely: "Nice to meet you."

Yet another student shrugs and says, "I don't have a problem with it."

Elsewhere in the video, Backholm receives the same sort of indulgent replies when alternately claiming to be seven years old, 6'5" tall, or Chinese. One student responds: "I mean, I might be a little surprised, but I would say, 'Good for you—like, yeah, be who you are!'"[1]

Be who you are.

Behold the faces, voices, and mindsets of today's Gen-Z, post-truth culture. Where reality is whatever you want it to be. Whatever you believe it to be. Whatever you *feel* it is, no matter how fanciful, far-fetched, or downright farcical it is on the face of it.

It's as if we've literally stepped into and are living out Hans Christian Andersen's famous fairy tale *The Emperor's New Clothes*.

In the story, you'll recall, two bogus weavers hoodwink the emperor into letting them create a magical new outfit for him. They claim it's made of materials that are "invisible to any one who was unfit for the office he held, or was incorrigibly stupid."[2]

In fact, the outfit is a complete fiction. But when it's "unveiled," no one—not even the emperor himself—dares to admit they can't see a thing. They play along with the fraud, for fear of being called stupid or unfit for office.

Nowadays, the expression *the emperor has no clothes* is used "to describe a situation in which people are afraid to criticize something or someone because the perceived wisdom of the masses is that the thing or person is good or important."[3]

Today, if you venture to challenge a post-truther's sincere fantasy or outright lie, prepare to be vilified. You'll be called every name in the book, accused of being judgmental, and probably blocked on a social media site or two. In short, you'll become a reviled nonperson—*canceled*, to use the term du jour.

What's driving this seeming madness? Partly, it's the supremacy of *feelings* and the deification of tolerance in today's post-truth, politically correct culture.

If one day I *feel* like being a 6'5" Chinese woman—even though quite obviously I'm not—who are you to correct or contradict me? It would hurt my feelings. It would smack of racism or misogyny. The gods of tolerance and kindheartedness demand that we *must* play along with whatever nonsense others put forth.

For the record, I believe that most people routinely play along with other people's fictions, so long as they aren't harmful to themselves or society. Refusing to do so, I agree, would be unnecessarily mean-spirited, even hateful.

As a child, I identified strongly with Superman. Since my family had very little money, I made my own Superman uniform—cape and all—out of my pajamas, underwear, socks, and a pillow case—all of which I colored using crayons. I ran around the house, jumping off beds, pretending to fly. Thankfully and understandably, neither my parents nor sisters challenged my fantasy.

But is there a limit to how far society can play along with the post-truthers' polite, tolerant game of make-believe before it becomes dangerous? Yes, of course.

In early 2020, a forty-five-year-old pedophile named Joseph Gobrick of Grand Rapids, Michigan, was tried for possessing pornographic images of young children being molested. His defense was that he sincerely self-identified as a little girl and was simply exercising First Amendment rights.

This was *his* reality.

"I've always been an eight-year-old girl," Gobrick insisted. "Even in my drawings and fantasies, I am always an eight-year-old girl."[4]

Fortunately for our society, the judge *did not* play along with Gobrick's dangerous delusion. Instead, he sent him to jail for ten to twenty years.

Another example of society grappling with post-truth behavior involves transgender individuals—especially biological male athletes who self-identify as females. These transgender athletes are hijacking women's sports, winning competitions hands down and blowing away decades-old sports records set by biological females.[5]

Transgender politics aside—and right now, the public debate is red hot—more and more states are urgently writing legislation intended to protect the integrity of women's sports. According to a recent Rasmussen poll, 54 percent of Americans favor such laws; only 28 percent oppose them.[6] For their part, transgender women are protesting that their civil rights are being violated.[7]

What about you? What's *your* reality?

What do you believe about yourself? Do you see yourself as younger than you are? Slimmer? Smarter?

Or, heaven forbid, do you see yourself as less beautiful than you are? Less worthy for a promotion? Less deserving of ever being happy?

Who among us lives free of all illusion?

Besides, certain illusions can be quite beneficial. An aspiring actor might see himself taking bows at the Academy Awards ceremony. An aspiring medical researcher might see herself getting the Nobel Prize for discovering a cure for cancer. Aspiring astronomers might see themselves discovering life on another planet.

Filling our minds with positive self-images and beliefs, even if they are fictional, can be a healthy way to encourage ourselves to do and be our best. They're not a bad thing.

But is there a limit to how much self-delusion we dare tolerate in ourselves before it endangers our physical, mental, emotional, and spiritual well-being—as well as our relationships with others and with the absolute truths of the universe? That is the all-important question we will explore in the remainder of the chapter.

WHAT IS YOUR WORLDVIEW?

Let me begin by asking you several questions intended to make you pause and look inward, to figure out what makes you tick. Yes, I know you're busy. But unless you take a few minutes to do this little exercise, you'll likely miss out on one or both of the two main purposes of this book: (1) to explain to you that absolute truth exists; and (2) to help you discover who and why you are.

My first question is this: *What is your worldview?* What are the basic elements of your reality? What do you believe about yourself, about others, about life in general? It's an important question because, as we've established, *believing is seeing.*

Your worldview, the sum total of your beliefs, determines everything about you. From how you see, think, react, and respond to things and people, to how you dress, speak, and carry yourself.

We have labels for all the major worldviews—Atheist, scientific, Christian, Muslim, Jewish, Buddhist, Hindu, Jain, Baha'i, Sikh, Zoroastrian, Shinto, Confucian, Taoist, and so forth—with Christianity, Islam, Hinduism, and Buddhism accounting for roughly three-fourths of the world's population.[8]

If you choose to adhere to a pre-labeled worldview, it's like you're buying something off the shelf. A *ready-to-wear* worldview, as it were—and there's nothing wrong with that.

But customization is both popular and possible nowadays. People routinely mix and match fundamentally compatible worldviews—such as the scientific and Christian worldviews, as I've done. It's called "pastiche" or "salad bar" spirituality.[9]

Many post-truthers—in the name of kindness, tolerance, and virtue signaling—have worldviews designed to be all things to all people. But inevitably their worldviews are internally contradictory. You can't honestly be a New Age pagan *and* a Christian, for example, even if you celebrate the winter solstice and Christmas Day with equal piety.

Whatever your worldview, here are three specific questions to ask yourself:

1. What's the *foundation* of your worldview?
2. What's the *size* of your worldview?
3. What's at the *center* of your worldview?

By answering these questions, you'll be dragging your worldview—which normally is embedded deep within your subconscious—out into the light of day. You'll be discovering, perhaps for the first time, *who* and *why* you are.

THE FOUNDATION OF YOUR WORLDVIEW

Throughout this book, I've presented you with plenty of evidence in support of this important thesis: Whatever your worldview, it is founded on *faith*. It is unavoidably founded on ideas and feelings that cannot be proved.

Even if your worldview is *evidence-based*, to use a popular buzzword, it is founded on axiomatic *beliefs* you can't prove, see, or even fully imagine. Your worldview, therefore, even if it is atheistic, is *faith based*.

I've also presented you with plenty of evidence for a second important thesis: Not all faiths are created equal. There's IQ-based faith, SQ-based faith, enlightened faith, and misguided faith. The best one of these—the one that an ideal worldview is founded upon—is *enlightened IQ-and-SQ-based faith*.

- It is the faith that, as I described in chapter 5, comes from putting on your 3D glasses—from looking at the world through the powerful, complementary lenses of your brain's IQ *and* SQ—and beholding the true width, breadth, depth, and meaning of reality.
- It is the faith that obeys IQ's strict rules and regulations *and* hearkens to SQ's subtle, subconscious voices and whispers.
- It is the faith that opens your eyes to both trivial, logical, IQ-based truths, realities, and experiences *and* profound, translogical, SQ-based truths, realities, and experiences.
- It is the faith that leads you, like a supernatural guide dog, to an expansive, hope-filled, life-affirming worldview founded solidly on enlightened beliefs that are consistent with the best available evidence.

THE SIZE OF YOUR WORLDVIEW

I began life with a rather small scientific-atheistic worldview, as I've already explained. It was only big enough to fit trivial truths and logical realities.

My diminutive worldview faced its first real test when I entered grad school and began diving into hard-core scientific ideas that positively defied logic: relativity (spatial and temporal), duality, the quantum vacuum, imaginary time, tachyons, virtual particles, black holes, white holes, worm holes, 4D spacetime, 10D strings, dark energy, dark matter, inflation, the multiverse, photons, quarks, and on and on.

I managed to squeeze these gigantic oddities into my tiny worldview—but not comfortably. Eventually it became an intellectual crisis, which set me off on a journey that took me far from the circumscribed scientific plantation where I had spent my life until then.

For decades I explored worldviews other than the strictly scientific one, until gradually and slowly, parsec by parsec, the diameter of my worldview grew. Finally, it expanded to the size it is today—capacious enough to accommodate faith not only in science's many far-out notions but also in the God of the Bible, in Jesus, and in the Holy Spirit.

Is your worldview as big as all that? Or is it the size that mine was when I was a scientific monk?

THE CENTER OF YOUR WORLDVIEW

Your worldview is your religion. Yes, I know, if you're not formally religious, you may not see it that way. But you *are* a religious person, even if you profess Atheism.

As we've seen, your worldview, whatever it might be, is founded on *faith*—on beliefs you can't prove, see, or fully imagine.

So the only remaining question is this: Which religion do you practice? In other words, *who* or *what* is at the *center* of your worldview?

Whatever or whoever is at the center, *that's* your god. It's the thing around which everything else in your life revolves. It's the thing you worship.

That bears repeating.

Whether it's the God of the Bible, the god of tolerance, the god of success, or whatever person or principle you hold most dear, that person or thing is your deity, the god at the center of your worldview, the definer of your religion.

Five hundred years ago, people argued about whether the earth or sun was at the center of the universe. It was a battle between a geocentric worldview and a heliocentric worldview.

We could ask the same question today: Is your worldview *geocentric* or *heliocentric*?

If you and your feelings are at the center of your universe, you have a geocentric worldview. It's small-minded because you can't imagine anything bigger or more important in the whole, wide world than you and your feelings.

For much of my life, that was exactly *my* worldview.

Growing up in East LA, all I wanted was to become a scientist. That dream was at the very center of my universe. Everything I did—everything I was—revolved around my goal of becoming a scientist. It was my god. That's why when I became a scientific monk at Cornell, I felt I'd hit the lottery. I was living my dream. Nothing and nobody else mattered to me. In fact, when it seemed my relationship with Laurel was getting serious, I told her straight out: "Even if we ever do get married, you'll always be second in my life to science."

Yikes! But that was my reality back then. Everything in my small-minded, self-centered life revolved around the tiny planet called Michael Jillion.

By contrast, people with heliocentric worldviews worship something other than themselves, their feelings, and their dreams. The scientific and Christian worldviews are examples of this because both urge us to cherish selflessness and absolute truth over self-centeredness and personal feelings.

YOUR TITANIC MOMENT

How does your worldview stack up to an *ideal* worldview—one that is founded on enlightened IQ-and-SQ-based faith? One that is big enough to accommodate both the universe *and* its Creator. One that is centered not on you and your feelings but on universal, eternal, absolute truths.

This is not an academic question. It's all-important how well *your* worldview compares to an ideal worldview because it determines how well you're prepared for a "Titanic Moment."

What's that? It's any crisis that threatens to break you. That threatens your peace of mind. That threatens your very life.

Like the one I faced at the bottom of the North Atlantic Ocean.

Like the one Laurel and I faced when trying to start a family.

Like the COVID-19 pandemic the entire world is battling as I write these words.

Titanic Moments reveal what you're really made of. Expose the weaknesses of your *worldview*.

When a Titanic Moment strikes—no matter how smart or successful you are, no matter how many powerful friends you have—you feel trapped and can't for the life of you figure out any escape. At that moment, your worldview is your most valuable possession.

Will it prove to be an *asset* . . . or a *liability*?

Remember, your worldview is the sum total of what you *believe*. What you believe will dictate how you *see* your predicament; and that, in turn, will dictate how you *react* to it.

Believing is seeing.

Seeing is reacting.

The engineers who designed the *Titanic* believed she was unsinkable. That belief defined their reality; consequently, they saw no need to stock the ship with lots of life vests and lifeboats. And they reacted accordingly.

If they hadn't believed the *Titanic* was unsinkable, it would've been a totally different story. Their misguided worldview contributed to the deaths of more than fifteen hundred people.

The passengers on the *Titanic*'s maiden voyage also believed the ship was unsinkable. Consequently, they saw no great risk in voyaging across the North Atlantic through the area known as Iceberg Alley.[10] And they reacted by shelling out big bucks for the privilege.

Believing is seeing.

Seeing is reacting.

If *your* reality is seriously at odds with *absolute reality*, when a Titanic Moment strikes, your worldview will sink your ship. Maybe you've gotten away with having a misguided worldview for years; but when disaster hits, your make-believe reality will blow up in your face.

For those unfortunate men, women, and children aboard the *Titanic*, such a moment arrived just before midnight on April 14, 1912, when their reality—belief in the unsinkability of the *Titanic*—collided with absolute reality in the form of a giant iceberg. In a matter of hours, their misguided faith cost them their lives.

Remember Kurt Gödel, the brilliant logician who proved the incompleteness theorems? During the rise of Nazi anti-intellectualism in Vienna in the 1930s, Gödel's good friend Professor Moritz Schlick was assassinated by a former student. When Gödel got the news, he began fearing for his own life. It was his Titanic Moment.

There was no real evidence that anyone wanted to assassinate him, but Gödel became obsessed with the belief that certain bad actors were out to poison him. Consequently, he would only eat food prepared by his wife, Adele.

After fleeing the Nazis in 1939, Gödel came to America and had a productive career at Princeton University's legendary Institute for Advanced Study. But his fears came with him. Later in life, when Adele was incapacitated for many months after suffering a stroke, Gödel stopped eating and eventually starved to death. His misguided worldview became his reality and cost him his life.

Believing is seeing.

Seeing is reacting.

So, once again, I ask you: What do you believe? What is your reality? What is your worldview? How big is it? And who or what is your god?

Please do not shrug off the effort. Because if it hasn't happened already, one day you *will* experience a Titanic Moment. When you do, your worldview, your beliefs, your god will spell the difference between hope and despair, courage and fear, wisdom and delusion. Above all, it will spell the difference between life and death.

YOUR DREAM

When I look back on my life, I recognize the importance of dreaming big. Ambitious dreams are powerful motivators—not only for the dreamer but for everyone in his or her orbit.

"I have a *dream* today!" exclaimed the Reverend Dr. Martin Luther King, Jr. in his 1963 speech to a crowd in front of the Lincoln Memorial in Washington, DC. "I have a dream that one day every valley shall be exalted, and every hill and mountain shall be made low, the rough places will be made plain, and the crooked places will be made straight.[11]

It's not likely Dr. King's enlightened dream will ever be fully realized, given our innate human imperfections. But it certainly mobilized a civil rights movement that to this day is radically transforming American society for the better.

My own dream of becoming a scientist propelled me from East Los Angeles to UCLA to Cornell to Harvard. But what if I'd stopped there? What if I'd only fulfilled my initial dream? Almost certainly I'd still be holed up in a lab somewhere—which I'm not suggesting would've been a totally bad thing.

But, in fact, I'm not just a scientific monk. And for one big reason.

During my single-minded drive to become a scientist, my geocentric worldview exploded into something much larger and less self-centered. It morphed into a heliocentric worldview that included not only truths I could see, prove, and imagine but truths I could *never* see, prove, or even imagine.

Because of that, I not only realized my childhood dream, but I discovered something far grander: *my destiny.*

By *destiny* I mean the unique purpose God has in mind for my life. It is preconceived by him but ultimately decided by me—by the day-to-day decisions I make in life.

The same is true for you.

A dream, you see, originates with *you*. A destiny, according to the Christian worldview, originates with *God*, your creator.

A dream might be big-hearted, well-intentioned, and even altruistic; but at its core it comes from an earthly place in a small, geocentric worldview. A destiny might involve poverty, humility, and even pain; but at its core it comes from a heavenly place in an expansive, heliocentric worldview.

A destiny is defined not solely by specific achievements but by personal character. In other words, a destiny is about *who* you become, not just *what* you become. In Christianity, the ideal character is like Christ's.

A destiny is not just a destination; it's a journey that rarely, if ever, follows a straight path. It's a journey with many surprising twists and turns, including exasperating dead ends and dashed hopes. And it's a journey that lasts an eternity, not just a lifetime.

A destiny is not always at odds with a dream. But often it is, and almost always, it is far grander and more breathtaking than a mere dream. The important thing is to recognize the fundamental difference between a destiny and a dream.

So how do you find your destiny? By intentionally seeking it out. In a moment I'll explain how, according to the Christian worldview, you can begin doing that.

For now, my point is this: Even though I can't *prove* that I have found my destiny—for no one can ever know with absolute certainty the mind and intentions of God—I believe with all my heart I have done so and am living it out.

Never in my entire life have I felt as thoroughly fulfilled and at peace with life as I do now—more so than even during those many contented days I spent as a scientific monk. What's more, judging from the messages I receive daily from strangers all over the world who read, listen to, or watch my communications, never have I made such a positive difference in other people's lives. And, finally, never before has my character been so well-rounded, empathetic, and selfless. And although I'm still far, far from perfect or where I want to be, I can see the character of Jesus being formed in my life.

All in all, even though I'm not twentysomething anymore, I feel totally energized—like a 1,000-horsepower sports car firing on all cylinders. Like an airplane that has found exactly the right altitude to have the jet stream at its back. Like an Olympic athlete operating in *the zone*—what Hungarian-American positive psychologist Mihaly Csikszentmihalyi calls *flow*,[12] what we physicists call *resonance*, and what I'm calling here your God-given *destiny*.

My destiny involved exploring the world's religions, seeking rigorous answers to deep questions that science and logic alone cannot answer.

My destiny involved meeting Laurel, reading the Bible with her, and marrying her. In the fall of 2021, we will celebrate our thirtieth wedding anniversary!

My destiny involved meeting Fred Graham, catching his attention with my explanation of a Foucault pendulum, and becoming a national television correspondent.

My destiny involved producing an award-winning family movie about generosity that introduced me to the extraordinary, selfless world of big-time philanthropy.

My destiny involves being a popular speaker who travels the world explaining the unmatched power of faith and its crucial importance to both science and Christianity.

My destiny involves writing books—including the one you're reading right now. A book in which I'm saying to you, *Your destiny is more awesome, more satisfying, more meaningful than even your wildest dreams!*

It's been widely reported that in 1968, Dr. King had a premonition that his days were numbered. On April 3, at the Mason Temple in Memphis, he gave a speech in which he said, "I've seen the Promised Land. I may not get there with you. But I want you to know tonight, that . . . I'm not worried about anything; I'm not fearing any man. Mine eyes have seen the glory of the coming of the Lord."[13]

The following day, he was assassinated at the Lorraine Motel.

I resonate strongly with Dr. King's seemingly prophetic words: "I'm not worried about anything." They cause me to flash back to my own Titanic Moment.

When staring death in the face at the bottom of the North Atlantic, I felt a similar, uncanny sense of peace. When my situation screamed confusion, alarm, and futility, my worldview filled me with wisdom, serenity, and courage.

Premonition or not, being assassinated was surely not part of Dr. King's dream. Yet, as a sincere Christian, he undoubtedly would see it as his destiny. He was destined to become a celebrated martyr for the ages, an everlasting inspiration to any and all who are oppressed.

FINDING YOUR DESTINY

When I spoke one time at George Mason University, a student told me she was going to college only to please her parents. It was their dream for her. She was desperately unhappy, she explained, but didn't know what to do about it.

I talked to her about the huge difference between a dream and a destiny. Afterward, she said, "Dr. Guillen, how do I find *my* destiny?"

"Well, here's how I found mine," I said to her. "Maybe it'll help you find yours."

I explained my experience in terms of two tuning forks. Tuning fork #1 is me. Tuning fork #2 is God.

When I strike *me*, I vibrate at a certain frequency. Years ago, that vibe was me madly pursuing my dream. It was me single-mindedly doing my own thing.

When I strike *God*, he vibrates, too. That's my God-given destiny calling.

All during my years as a scientific monk, those two frequencies—my dream and my destiny—were very different. They were, as we say in physics, very *dissonant*.

In physics, we can measure the exact dissonance between two frequencies by listening for undulating hums we call *beats*. The greater the dissonance, the greater the number of beats or undulating hums per second.

Dissonance in our lives usually appears in the form of dissatisfaction, discouragement, dejection, or maybe even suicidal thoughts.

For me as a grad student at Cornell, the dissonance appeared as an intellectual dissatisfaction with science's answer to my big question: *How did this amazing, mostly invisible universe of ours come to be so amazing and mostly invisible?* The standard cosmological model had too many problems to be completely believed.

That dissonance ultimately drove me to explore other disciplines, other worldviews. It opened my mind to a bigger universe than the one I was studying in my physics, astronomy, and math classes. It opened my mind to the possibility that there was more to life than my basement laboratory. It opened my mind to the possibility that there was something even greater than a dream.

Gradually, I came to *believe* I had a destiny. And afterward, little by little, I began to *see* that destiny. And finally, I *reacted* by giving myself over to it wholeheartedly, even though it meant leaving behind my beloved lab.

Believing is seeing.

Seeing is reacting.

By now, I trust you're catching on to this all-important algorithm.

Now, think again about those two tuning forks.

When I was a scientific monk, *my* tuning fork and *God's* tuning fork were very far apart—not only in frequency but also in spacetime. His voice was calling out—as it always is—but I was too far away to hear it. So I kept singing my own tune.

Exploring the world's religions—becoming more open-minded about the possibility that a god exists—gradually brought the two tuning forks closer together.

In physics, we know that when two tuning forks—one that has been struck and one that is still—are brought really close together, something amazing happens: The quiescent, non-vibrating fork picks up on the other's vibrations and begins vibrating at exactly the same frequency! We call this *sympathetic vibration.*

As I told the student at George Mason, if she wished to increase the odds of finding her destiny, she had to do at least two things: *get close to God* and *become still.*

Doing this, I hastened to add, didn't guarantee that she'd find her destiny—not right away, anyway. But *not* doing these two things did guarantee she'd never find it.

In math, we call such criteria "necessary but not sufficient."

They're not sufficient because there could be issues in our lives that effectively jam or sabotage our communication with God. In physics we call it a *damping phenomenon.*

But the two criteria are necessary, and for exceptionally good reasons.

The first one—drawing close to God—is necessary because he is the *author* of your destiny. Now, on the chance you believe that God doesn't even exist, start by opening your mind to the possibility that he does. Because unless you're at least willing to *believe* that much, you'll never be motivated to *see* whether he's real or not—to do the hard work it takes to find out. And, therefore, you'll *react* by simply continuing to live as though God is a fiction, and you'll die without ever discovering the truth for yourself.

Second, you must silence yourself and become still. Set aside your own ambitions and open yourself up to *all* possibilities. In the Christian worldview, this discipline is what is meant by loving God with all your heart, mind, soul, and strength, and surrendering yourself to his will—that is, to his *destiny* for you.

The Bible puts it this way: "Be still, and know that I am God!"[14]

In his first letter to the church in Corinth, Paul the apostle explains that when we "surrender" to God, our entire being is taken over by his Holy Spirit, which guides us step-by-step and day-by-day toward our God-given destiny.

> No eye has seen, no ear has heard,
> and no mind has imagined
> what God has prepared
> for those who love him.[15]

Are you picking up on the Bible's translogical vibe here? It's speaking about truths, realities, and experiences that we cannot see, prove, or even imagine.

Paul goes on to describe all the remarkable things that happen when we get close to God, quiet ourselves, and start listening to his all-knowing, all-seeing Holy Spirit.

> It was to us that God revealed these things *by his Spirit*. For his Spirit searches out everything and shows us God's deep secrets. No one can know a person's thoughts except that person's own spirit, and no one can know God's thoughts except God's own Spirit. And we have received God's Spirit (not the world's spirit), so we can know the wonderful things God has freely given us.
>
> When we tell you these things, we do not use words that come from human wisdom. Instead, we speak words given to us by the Spirit, using the Spirit's words to explain spiritual truths. But people who aren't spiritual can't receive these truths from God's Spirit. It all sounds foolish to them and they can't understand it, for only those who are spiritual can understand what the Spirit means.
>
> Those who are spiritual can evaluate all things, but they themselves cannot be evaluated by others.
>
> For,
>
> "Who can know the LORD's thoughts?
> Who knows enough to teach him?"
>
> But we [who have surrendered to God's Holy Spirit] understand these things, for we have the mind of Christ.[16]

As I've explained, an epiphany happened to me in the 1990s. For the first time in my dream-driven life, I began listening for and picking up on God's voice—and then following its lead. I began vibrating sympathetically with my destiny.

From then on, there was no looking back.

I abandoned my lab but not my dream—far from it. For my destiny *included* my dream and so much more.

My destiny took me to places that the little Mexican kid from East LA had never even dreamed of. It took him far, far, far beyond what his eyes had ever seen, his ears had ever heard, and his mind had ever imagined—exactly as the Bible claims.

And now, my fellow traveler, by reading this book, *your* journey has joined with mine. Now and forever, you are part of my destiny, and I of yours.

What are you going to do about it?

Are you merely chasing a dream? Or are you seeking after your destiny?

Are you facing a Titanic Moment? If so, how well is your reality faring against absolute reality?

The answers depend on what you believe. They depend on your worldview.

So one last time: *What is your worldview?*

Is it big enough to encompass your destiny? Or is it so small that you believe only in things you can see, prove, and imagine?

Is it big enough to allow for the possible existence of things you cannot see, prove, or imagine? Or do you live by the small-minded, logical, materialistic motto of "Seeing is believing"?

If so, I'm inviting you to live by the cosmic-minded, translogical, metaphysical principle: *Believing is seeing.*

Research shows that most people lock and load their worldview by age twenty-six. So don't put this off. And don't let anyone make the choice for you, not even me. Above all, make certain it's a choice based on enlightened IQ-and-SQ-based faith; and on the best, most reliable evidence available.

Do your homework, as I did. Study the various worldviews, including the scientific and atheistic ones. Ask questions. Demand honest, rigorous answers.

Finally, I wrote this book not to defend a particular worldview but to explain how and why I chose mine.

I didn't even write this book to defend absolute truth. As a popular saying goes: Truth needs no defense. Truth is like a lion. Simply turn it loose and it will defend itself.

That's been my purpose here—to turn the truth loose. Now it's up to you to figure out what to do with it.

13

PUTTING YOUR FAITH TO THE TEST

Faith is the bird that feels the light and sings
when the dawn is still dark.

RABINDRANATH TAGORE

YOU are a person of faith. Everyone is. Whether you realize it, or wish to admit it, your worldview is founded on having faith in things that cannot be seen, proven, or even fully imagined. It's what I've endeavored to explain in this book.

Is your worldview supported by *misguided faith* or *enlightened faith*? How big is it? Who or what is your god? This brief test will help give you a sense of how your worldview stacks up against an ideal worldview, which I explained in the previous chapter.

When you've completed the test, compare your responses to the answer key on page 228. Give yourself a point for every matching response. Then add up the points to get your **TOTAL SCORE**.

Next, multiply your total score by five to get your **TRUTH SCORE**, which rates how well *your* reality lines up with *absolute reality*. Think of it as a reality check on your worldview.

Suppose your total score is 14. Multiply that number by 5, and you get a truth score of 70. It means *70 percent* of your reality lines up with absolute reality.

Next, subtract your truth score from 125 to get your **TITANIC SCORE**, which rates how likely it is that your worldview will sink your ship when faced with a Titanic Moment.[1]

For example, use the aforementioned truth score of 70 percent. Subtract 70 from 125 and you get 55. It means that when faced with a Titanic Moment, you'll have about a 55 percent chance of making a *misinformed* decision that'll harm you.

If you've read this book carefully and understand the material, you should score well. If you don't score well, then for your edification the **ANSWER KEY** includes the page numbers where I explain the topics that inspired the test questions.

One last thing: Answer the questions honestly, not in the way you think they should be answered.

Good luck!

TEST QUESTIONS

1. *Objective, universal, absolute truth does not exist.*
   True __________ False __________

2. *Faith has no role in hard-core logic and science.*
   True __________ False __________

3. *Space and time are perfect examples of subjective reality.*
   True __________ False __________

4. *There is nothing really unique about humans; we're basically like any other animal species.*
   True __________ False __________

5. *Mathematicians have shown it is possible to prove that God exists.*
   True __________ False __________

6. *"Seeing is believing" is a much wiser way to live than "believing is seeing."*
   True __________ False __________

7. *Science has no need for God.*
   True __________ False __________

8. *Faith is a potent, natural ability.*
   True __________ False __________

9. *Science and Atheism are more intellectually compatible than science and Christianity.*
   True __________ False __________

10. *Sometimes opposites can both be true.*
    True __________ False __________

11. *A wise person should never doubt the credibility of a scientific consensus.*
    True __________ False __________

12. *If something is not logical, it's nonsense and should not be believed.*
    True __________ False __________

13. *Our telescopes are now powerful enough to see 95 percent of what's out there in the universe.*
    True __________ False __________

14. *The more science learns about the universe, the more supernatural and mysterious it appears.*
True __________ False __________

15. *Science has pretty much proven that life on Earth began when molecules assembled themselves into a simple, self-replicating organism.*
True __________ False __________

16. *There is credible evidence that the universe was designed for life.*
True __________ False __________

17. *Faith is a crutch for people who can't think logically and can't stomach the truth.*
True __________ False __________

18. *There are absolute truths that cannot be proved.*
True __________ False __________

19. *Scientific experiments are objective ways of getting at the truth.*
True __________ False __________

20. *You and I are logical paradoxes.*
True __________ False __________

TALLY YOUR POINTS
1.
2.
3.
4.
5.
6.
7.
8.
9.
10.
11.
12.
13.
14.
15.
16.
17.
18.
19.
20.
Total Score ____________
____________________ x 5 = ____________________ percent *Total Score* *Truth Score*
125 minus ________________ = ____________________ percent *Truth Score* *Titanic Score*

ANSWER KEY

If you want to see where in the book a question is discussed, see the page numbers in parentheses.

1. False (30-31, 37-40, 132-135)
2. False (ix, xi-xii, 71-72, 84-87, 89-103)
3. True (15, 31, 50, 132)
4. False (41-42, 45, 58, 60, 102-103)
5. False (108-114)
6. False (9, 38-39, 80-81, 91)
7. True (60-62)
8. True (72, 86-87)
9. False (29-40)
10. True (24-27, 48)
11. False (108-114, 130)
12. False (25-27)
13. False (9, 15, 140)
14. True (61-62)
15. False (148-154)
16. True (33-35)
17. False (xi-xii, 71-72, 75-87)
18. True (30-33, 37-39)
19. False (97-102, 127-129)
20. True (108-114, 125)

APPENDIX A

THE COSMIC DISTANCE LADDER

1. Super-Short-Range Rungs (up to six hundred million miles): Laser, radar, radio

These rungs estimate distance by measuring the time it takes electromagnetic signals to reach us, given they travel at 186,000 miles per second.

In 1969, moon-walking Apollo astronauts left behind a special mirror on the lunar surface.[2] A laser beam travels from Earth to the moon, reflects off the mirror, and returns. By timing the round trip, we can calculate the distance to the moon within *one inch*. On average, the moon is 238,855 miles away from us.[3]

Likewise, we shoot radar signals to large, nearby planets and moons and time how long it takes the signals to travel there, bounce off, and return. That tells us how far those bodies are. What's more, the telltale distortions in the reflected radar signals tell us things about a body's spin, topography, and even atmosphere.

It's the same story with spacecraft we launch into deep space. The time it takes for its radio signal to reach us tells us how far away it is. And the distortions in the signal tell us something about what it might have passed through in getting to us—like a planet's rings or its atmosphere.

2. Short-Range Rung (up to 30,000 ly): Parallax

This rung estimates distance by measuring a well-known optical illusion called *parallax*. Hold your finger up to your nose and look at it with one eye at a time: first right, then left. Do you see what happens? The finger appears to jump—left to right—even though it isn't actually moving.

That's parallax. An object's position appears to move when you shift your perspective. The size of the shift depends on the distance of the object from you and the distance between your two perspectives.

The *bigger* the distance between you and the object, the *smaller* the parallax. You can see this for yourself by moving your finger toward and away from your nose, measuring the parallax each step of the way.

Also, the *bigger* the distance between your two perspectives, the *bigger* the parallax. Have someone stand in front of you, then look at them with just one eye. Now (with one eye still closed) move sideways and notice how the person's position against the background appears to shift with your sideways movement.

In astronomy, a telescope plays the role of a single eye. To simulate having two eyes, it might, for example, look at a star at two different times of year (when Earth is at opposite points in its orbit). That makes for a big distance between *perspectives*, which makes for a big parallax. The size of the parallax tells us how far away the star is from us.

Presently, our best instrument for measuring parallax is the European Space Agency's Gaia satellite. As of this writing, the orbiting telescope has reliably measured the parallaxes of—and therefore the distances to—more than one billion stars. The farthest ones are some 30,000 light-years away.[4]

3. Medium-Range Rung (up to 200,000 ly): Ordinary stars

This rung estimates distance by exploiting a relationship among the three essential traits of any ordinary star: *color*, *temperature*, and *true brightness* (absolute magnitude).

Using a spectroscope—a prism-like device that analyzes starlight—we can tell a star's main color. That, in turn, tells us the star's temperature and absolute magnitude.

A *blue* star burns hotter and more brightly has a greater absolute magnitude than a *red* star. It's the same on your gas stove: A blue flame is hotter and brighter than a red flame.

This is not an exact science, mind you. Only by actually traveling to a star could we honestly measure its absolute magnitude. But there's plenty of evidence to support this relationship between color, temperature, and absolute magnitude.

There's another thing we need to know to understand how this rung works.

A light bulb naturally looks very bright if we bring it right up to our nose. But it looks less and less bright as we move it away.

Scientists have discovered that the light bulb's brightness decreases by a factor of four when its distance from you doubles. (Note: One-fourth is equal to the inverse of two squared.) We call this predictable behavior the *inverse-square law*.

This rung mashes the inverse-square law with the color-temperature-brightness relationship.

Suppose we peer out into deep space with a telescope and spot a garden-variety green star. From its green color, we immediately know its absolute magnitude.

Next, still peering through the telescope, we use a light meter to measure the green star's *apparent magnitude* (i.e., how bright it *appears* to us here on Earth). Its apparent magnitude will always be smaller than its absolute magnitude because of the star's enormous distance from us.

Finally, using the inverse-square law, we can explain its diminished brightness by estimating the distance the starlight must have traveled to reach us. That distance is directly related to the difference between the star's *absolute* magnitude and its *apparent* magnitude.

Using this technique, we can estimate the distances of all the ordinary stars in our Milky Way galaxy—and even a bit beyond. This includes all stars as far away as about 200,000 light-years.

4. Long-Range Rung (up to ten million ly): Unusual stars

This rung estimates distance by exploiting the discovery of pulsating stars called Cepheid variables. Each Cepheid blinks at a rate that depends on its absolute magnitude. The faster the star blinks, the smaller its absolute magnitude.

Altogether, this rung requires that we follow these three steps:

1. Look through a telescope and time the Cepheid's rate of blinking. That tells us its absolute magnitude.
2. Still looking through the telescope, use a light meter to measure the star's apparent magnitude—how bright it appears to us.
3. Use the aforementioned inverse-square law to estimate how many miles the starlight must have traveled to diminish in brightness. The mileage is directly related to the difference between the star's absolute magnitude and its apparent magnitude.

Using this technique, we can estimate the distances to galaxies well beyond the Milky Way—out to about ten million light-years.

5. Super-Long-Range Rungs (up to ten billion ly): Galaxies, supernovae, redshifts

These final rungs estimate distance by making informed assumptions about the absolute magnitudes of certain very distant objects. I'll mention four of the techniques.

The first one involves *spiral galaxies*. These giant pinwheels, including our Milky Way, spin once every billion years or so. Their exact spin-rate is directly related to their absolute magnitude. The faster a spiral galaxy spins, the greater its absolute magnitude. We call this the *Tully-Fisher relation*.

By clocking a spiral galaxy's spin-rate, therefore, we know its absolute magnitude. Which, when compared with its apparent magnitude, tells us its approximate distance using, once again, the inverse-square law.

The second technique involves *elliptical galaxies*. These football-shaped galaxies have stars that move in all directions, like cars in a demolition derby. The *diversity* of these speeds correlates with the galaxy's absolute magnitude. The greater the diversity, the larger the elliptical galaxy's absolute magnitude. We call this the *Faber–Jackson relation*.

By auditing the speeds of stars within an elliptical galaxy, we can surmise its absolute magnitude. Which, when compared with its apparent magnitude, tells us its approximate distance using the inverse-square law.

The third technique involves *supernovae*, dying stars that exit the world in a blaze of glory. The details of a supernova's final performance are directly related to its absolute magnitude.

By carefully observing a supernova's death throes, we can surmise its absolute magnitude. Which, when compared with its apparent magnitude, tells us its approximate distance using the inverse-square law.

The fourth technique involves an epic discovery by American astronomer Edwin Hubble. Nearly a century ago, he discovered that galaxies are flying away from each other, as if they were shrapnel from a massive explosion. It led to today's big bang theory and all its variations.

Hubble found that a galaxy's getaway speed is directly related to its distance from us. The farther the distance, the faster the speed. It's called, appropriately enough, *Hubble's Law*. By measuring a galaxy's getaway speed, Hubble's Law tells us the galaxy's distance from us.

These four techniques represent our best way of estimating distances—from close-by all the way out to the far side of the universe. To regions of space that are ten billion light-years away from us. That's *six hundred billion billion miles* from planet Earth!

APPENDIX B

WORLDVIEWS

All worldviews are faith-based. But there are different kinds of faith, and not all of them are created equal. Some are more reliable, prescient, and constructive than others.

One of the most important questions you can ask yourself in life is this: *What is my worldview?*

Your worldview *defines* you.

When things are going well, your worldview determines how perceptive, successful, and joyful you can be. When things go south, your worldview determines whether you sink or soar.

There are twelve basic types of worldview, based on their foundation, size, and center—which you will understand having read this book. A typical person doesn't have any one worldview in its pure form but an alloy of two or more of the archetypes.

WORLDVIEW ARCHETYPES

The best of all worldviews is *enlightened IQ-and-SQ-based heliocentric*. It's the largest, most selfless worldview possible and is solidly founded on absolute truth—on the best available evidence.

The worst of all worldviews is *misguided IQ-based geocentric*. It's the tiniest, most self-centered worldview possible and is based on feelings, not absolute truth—not on the best available evidence.

Here, then, are the twelve archetypal worldviews, listed from most desirable to least desirable. Look at each one carefully, and honestly answer the all-important question: *What is my worldview?* Nothing less than *your life*—in good times and bad—depends on which type of worldview you have.

ENLIGHTENED IQ-AND-SQ-BASED HELIOCENTRIC

- Based on absolute truth; on the best available evidence
- Values logical reasoning
- Values translogical revelation
- Selfless

ENLIGHTENED IQ-AND-SQ-BASED GEOCENTRIC

- Based on absolute truth; on the best available evidence
- Values logical reasoning
- Values translogical revelation
- Self-centered

ENLIGHTENED SQ-BASED HELIOCENTRIC

- Based on absolute truth; on the best available evidence
- De-emphasizes logical reasoning
- Emphasizes translogical revelation
- Selfless

ENLIGHTENED SQ-BASED GEOCENTRIC

- Based on absolute truth; on the best available evidence
- De-emphasizes logical reasoning
- Emphasizes translogical revelation
- Self-centered

ENLIGHTENED IQ-BASED HELIOCENTRIC

- Based on absolute truth; on the best available evidence
- Emphasizes logical reasoning
- De-emphasizes translogical revelation
- Selfless

ENLIGHTENED IQ-BASED GEOCENTRIC

- Based on absolute truth; on the best available evidence
- Emphasizes logical reasoning
- De-emphasizes translogical revelation
- Self-centered

MISGUIDED IQ-AND-SQ-BASED HELIOCENTRIC

- Based on feelings, not evidence; doesn't believe in absolute truth
- Values logical reasoning
- Values translogical revelation
- Selfless

MISGUIDED IQ-AND-SQ-BASED GEOCENTRIC

- Based on feelings, not evidence; doesn't believe in absolute truth
- Values logical reasoning
- Values translogical revelation
- Self-centered

MISGUIDED SQ-BASED HELIOCENTRIC

- Based on feelings, not evidence; doesn't believe in absolute truth
- De-emphasizes logical reasoning
- Emphasizes translogical revelation
- Selfless

MISGUIDED SQ-BASED GEOCENTRIC

- Based on feelings, not evidence; doesn't believe in absolute truth
- De-emphasizes logical reasoning
- Emphasizes translogical revelation
- Self-centered

MISGUIDED IQ-BASED HELIOCENTRIC

- Based on feelings, not evidence; doesn't believe in absolute truth
- Emphasizes logical reasoning
- De-emphasizes translogical revelation
- Selfless

MISGUIDED IQ-BASED GEOCENTRIC

- Based on feelings, not evidence; doesn't believe in absolute truth
- Emphasizes logical reasoning
- De-emphasizes translogical revelation
- Self-centered

NOTES

INTRODUCTION: WHY I WROTE THIS BOOK

1. *PISA 2018 Results: What Students Know and Can Do*, vol. I (Paris: OECD Publishing, 2019), 3, 15, 87.

CHAPTER 1: CALIFORNIA DREAMIN' . . . AND BEYOND

1. *Frankenstein*, directed by James Whale (Universal Pictures, 1931).
2. Translated, *Concilio Latino Americano de Iglesias Cristianas* is the Latin-American Council of Christian Churches. Its current president is the Reverend José Ángel Nuñez.
3. The LNS is now called the Laboratory for Elementary-Particle Physics.
4. Jackson Ryan, "Scientists Propose a New Type of Dark Matter and How We Can Find It," CNET, June 11, 2019, https://www.cnet.com/news/scientists-propose-a-new-type-of-dark-matter-and-how-we-can-find-it/.
5. For a touching and well-deserved tribute to Professor Liboff, see H. Roger Segelken, "Physicist Richard L. Liboff Dies At 82," *Cornell Chronicle*, May 22, 2014, https://news.cornell.edu/stories/2014/05/physicist-richard-l-liboff-dies-82.
6. See Michael A. Guillen and Richard L. Liboff, "Kinetic Theory of Galaxy Correlations," *Monthly Notices of the Royal Astronomical Society* 231, no.4 (April 1988): 957–968, https://doi.org/10.1093/mnras/231.4.957.
7. In the midst of all these events, a local TV producer chose me to host a one-hour prime-time special called *Time, Tides & Tuning Forks*. It received monster ratings and won an Emmy Award. Those were heady days indeed.

CHAPTER 2: THE AWAKENING

1. See chapter 10, "Having Faith in Astronomy."
2. For more about the works of Hermann Hesse, see Marianna Hunt, "The Best Books by Hermann Hesse You Should Read," Culture Trip, October 4, 2016, https://theculturetrip.com/europe/switzerland/articles/the-best-books-by-hermann-hesse-you-should-read/.

3. For more on this, see chapter 11, "Having Faith in God."
4. Malachi 4:6, NIV.
5. Stefan Rozental, ed., *Niels Bohr: His Life and Work as Seen by His Friends and Colleagues* (New York: Wiley, 1967), 328.
6. Max Delbrück and Gunther S. Stent, *Mind from Matter? An Essay on Evolutionary Epistemology* (Palo Alto, CA: Blackwell Scientific Publications, 1986), 167.
7. See chapter 9, "Having Faith in Physics."

CHAPTER 3: THE RECKONING: PART 1

1. Albert Einstein, *The Collected Papers of Albert Einstein*, trans. Albert Engel, vol. 7, *The Berlin Years: Writings, 1918-1921 (English translation supplement)* (Princeton, New Jersey: Princeton University Press, 2002), 109, emphasis in the original.
2. Gödel proved *two* incompleteness theorems. I speak here only of the first one. If you want to know more about both, see chapter 8, "Having Faith in Mathematics." See also Panu Raatikainen, "Gödel's Incompleteness Theorems," *Stanford Encyclopedia of Philosophy*, April 2, 2020, https://plato.stanford.edu/entries/goedel-incompleteness/.
3. See chapter 2, "The Awakening."
4. Martin Rees, interview with Ian Sample, *The Guardian*, April 6, 2011, https://www.theguardian.com/science/2011/apr/06/astronomer-royal-martin-rees-interview.
5. Andrei Linde, quoted in Tim Folger, "Science's Alternative to an Intelligent Creator: The Multiverse Theory," *Discover*, November 10, 2008, https://www.discovermagazine.com/the-sciences/sciences-alternative-to-an-intelligent-creator-the-multiverse-theory.
6. There are, in fact, many more *just right* vital signs in the universe than the six in Sir Martin's book. See, for example, Klaas Landsman, "The Fine-Tuning Argument: Exploring the Improbability of Our Existence," in *The Challenge of Chance: The Frontiers Collection*, ed. Klaas Landsman and Ellen van Wolde (Springer, Cham: 2016), https://link.springer.com/chapter/10.1007/978-3-319-26300-7_6.
7. The exact odds vary because they depend on how many people enter the lottery. Generally speaking, the odds are one in hundreds of millions.
8. Sean M. Carroll, quoted in John Johnson, Jr., "Studying Time's Mysteries, and the Multiverse," *Los Angeles Times*, June 28, 2008, https://www.latimes.com/archives/la-xpm-2008-jun-28-sci-carroll28-story.html.
9. Lee Smolin, "You Think There's a Multiverse? Get Real," *New Scientist*, no. 3004 (January 17, 2015): 24–25.
10. Sabine Hossenfelder, "Post-Empirical Science Is an Oxymoron," *BackRe(action)* (blog), July 12, 2014, http://backreaction.blogspot.com/2014/07/post-empirical-science-is-oxymoron.html.
11. "Martin Rees - Why Cosmic Fine-Tuning Demands Explanation," Closer To Truth, YouTube video, 5:59, January 23, 2017, https://youtu.be/E0zdXj6fSGY?t=221.
12. Quoting Sidney Aaron in "The Final Truth of All Things," *Altered States*, directed by Ken Russell (Warner Brothers, 1980), https://www.imdb.com/title/tt0080360/quotes/qt0187612.
13. "Feminist MAGA Hat Meltdown at Women's March LA," FOG CITY MIDGE, YouTube video, 2:51, January 25, 2019, https://youtu.be/SnoCyCBv8v8?t=91.
14. "Thomas A. Edison on Immortality: The Great Inventor Declares Immortality of the Soul Improbable," interview with Edward Marshall, *Columbian magazine*, vol. 3, no. 4 (January 1911).

15. Steven Weinberg, *The First Three Minutes: A Modern View of the Origin of the Universe* (New York: Basic Books, 1993), 154.
16. John 18:38.
17. John 14:6-7, NIV.
18. Isaiah 44:6.
19. 1 Timothy 2:5.
20. John 20:25, NIV.
21. John 20:27.
22. John 20:28.
23. John 20:29.
24. Genesis 1:1.

CHAPTER 4: THE RECKONING: PART 2

1. See, for example, Bert Thompson, "The Origin, Nature, and Destiny of the Soul [Part II]," *Reason & Revelation* (newsletter), March 2000, http://apologeticspress.org/pub_rar/20_3/0003.pdf.
2. Genesis 1:27.
3. See, for example, Alyson Kalhagen, "Facts About African Grey Parrots," The Spruce Pets, January 13, 2020, https://www.thesprucepets.com/facts-about-african-grey-parrots-390715.
4. The proper nomenclature is a matter of some spirited debate. If you allow Neanderthals to come under the umbrella of *Homo sapiens*—i.e., *Homo sapiens neanderthalensis*—then anatomically and behaviorally modern humans are classified as *Homo sapiens sapiens*.
5. Sarah Wurz, "The Transition to Modern Behavior," The Nature Education Knowledge Project, 2012, https://www.nature.com/scitable/knowledge/library/the-transition-to-modern-behavior-86614339/.
6. Jared Diamond, "The Great Leap Forward," in Linda S. Hjorth, Barbara A. Eichler, Ahmed S. Khan, and John A. Morello, *Technology and Society: Issues for the 21st Century and Beyond*, 3rd ed. (New York, Pearson, 2007), 15.
7. Diamond, "The Great Leap Forward," 15–16.
8. "Howard Gardner on Multiple Intelligences," Edutopia, YouTube video, 7:54, July 9, 2010, https://youtu.be/iYgO8jZTFuQ?t=73.
9. For a readable description and critique of Gardner's multiple intelligences, see Kendra Cherry, "Gardner's Theory of Multiple Intelligences," Verywell Mind, July 17, 2019, https://www.verywellmind.com/gardners-theory-of-multiple-intelligences-2795161.
10. Howard E. Gardner, *Frames of Mind: The Theory of Multiple Intelligences* (New York: Basic Books, 2011), xiv.
11. 1 John 1:5, NIV.
12. The phenomenon of cancelling and reinforcing waves is called wave interference. The pattern they create on the screen is called an interference pattern.
13. Hertz's experiment is called the photoelectric experiment.
14. See "The Nobel Prize in Physics 1921," Nobel Prize, https://www.nobelprize.org/prizes/physics/1921/summary/. Interestingly, Einstein never won a Nobel Prize for his two theories of relativity, even though they are the discoveries that most laypeople associate with him.
15. It's also called *pair creation* or *pair production*.

16. It's also called *pair annihilation.*
17. Jesus said, "I and the Father are one," John 10:30, NIV.
18. 1 John 1:5, NIV.
19. John 1:14, NIV.
20. 1 Corinthians 15:42, GNT.
21. Philippians 3:21.
22. Isaiah 45:5, NIV.
23. Exodus 3:14.

CHAPTER 5: THE RESULT

1. Guillen, *Can a Smart Person,* 150.
2. 1 Corinthians 13:12, WEY.
3. "Ipsos Global @dvisory: Supreme Being(s), the Afterlife and Evolution," Ipsos, April 24, 2011, https://www.ipsos.com/en-us/news-polls/ipsos-global-dvisory-supreme-beings-afterlife-and-evolution.
4. Anthony F. C. Wallace, *Religion: An Anthropological View* (New York: Random House, 1966), 264–265.
5. Michael Lipka and David McClendon, "Why People with No Religion Are Projected to Decline as a Share of the World's Population," Pew Research Center, April 7, 2017, https://www.pewresearch.org/fact-tank/2017/04/07/why-people-with-no-religion-are-projected-to-decline-as-a-share-of-the-worlds-population/.
6. Lipka and McClendon, "Why People with No Religion."
7. Sam Harris, "The Problem with Atheism," *Sam Harris* (blog), October 2, 2007, https://samharris.org/the-problem-with-atheism/.
8. France was the only other country with such deep-water submersibles.
9. The shininess of the propeller really caught my eye. I wondered, *How could it have such a polished appearance, given how long it had been submerged in icy seawater?* No one I've talked to about it can explain it. I'm guessing it was the result of some kind of electrolytic reaction.
10. Psalm 139:7-10, ESV, emphasis added.
11. John 3:16, NIV, emphasis added.
12. See chapter 7, "Having Faith in the Scientific Method."
13. Mark 5:34.

CHAPTER 6: HAVING FAITH IN FAITH

1. H. L. Mencken, *Prejudices: Third Series* (New York: Alfred A. Knopf, 1922), 267–268.
2. The syllogism is consistent with the best available evidence; that is, we know of nothing in nature that contradicts either the axiom (which is the mathematical definition of evenness) or the conclusion.
3. Remember, when something is not logical, it can be outright illogical (i.e., nonsensical) or it can be translogical (i.e., profound).
4. *The Crown,* "Moondust," Netflix video, 56:26, November 17, 2019, https://www.netflix.com/watch/80215737. Quoted segment begins at 50:01.
5. See "Brain Bank," *National Geographic* video, 3:08, https://video.nationalgeographic.com/video/00000144-0a23-d3cb-a96c-7b2f8ae30000.
6. See, for example, an interview with psychiatrist Iain McGilchrist, "Certainty and

Flow, Iain McGilchrist (Part 1 of 2)," Rebel Wisdom, YouTube video, 37:29, December 17, 2018, https://www.youtube.com/watch?v=fI1ngqwH5us.
7. Richard Ivry, quoted in David Wolman, "The Split Brain: A Tale of Two Halves," *Nature*, March 14, 2012, https://www.nature.com/news/the-split-brain-a-tale-of-two-halves-1.10213.
8. *Corpora callosa* is the plural form of *corpus callosum.*
9. "The Nobel Prize in Physiology or Medicine 1981," Nobel Prize, https://www.nobelprize.org/prizes/medicine/1981/summary/.
10. See "Split brain behavioral experiments," Neuroslicer, YouTube video, 4:35, April 18, 2007, https://www.youtube.com/watch?v=ZMLzP1VCANo.
11. See chapter 2, "The Awakening"; chapter 4, "The Reckoning: Part 2"; and chapter 5, "The Result."
12. Edward Young, "The Relapse," *Night Thoughts*, line 176 (London: T. Heptinstall, 1798), 93.
13. Iain McGilchrist, *The Master and His Emissary: The Divided Brain and the Making of the Western World* (New Haven, CT: Yale University Press, 2009), 18.
14. Wolman, "The Split Brain."
15. "Early Split Brain Research Gazzaniga," veronchiquita, YouTube video, 10:43, May 19, 2007, https://youtu.be/0lmfxQ-HK7Y?t=600.
16. John William Draper, *History of the Conflict between Religion and Science* (London: Henry S. King & Co, 1875), vi.
17. The age-old synergy between science and religion has been documented in many well-researched and well-written books by highly respected expert authors. See, for example, James Hannam, *The Genesis of Science: How the Christian Middle Ages Launched the Scientific Revolution* (Washington, DC: Regnery, 2011).
18. Albert Einstein, *Ideas and Opinions* (New York: Dell, 1973), 54–55.
19. Ronald W. Clark, *Einstein: The Life and Times* (New York: World, 1971), 425.
20. Walter Isaacson, *Einstein: His Life and Universe* (New York: Simon and Schuster, 2008), 390.
21. Bob Samples, *The Metaphoric Mind: A Celebration of Creative Consciousness* (Reading, MA: Addison-Wesley, 1976), 26.
22. Samples, *The Metaphoric Mind*, 26.
23. *A Rumor of Angels*, directed by Peter O'Fallon (Cinetel Films, 2000).

CHAPTER 7: HAVING FAITH IN THE SCIENTIFIC METHOD

1. Ivan Petrovich Pavlov, *Lectures on Conditioned Reflexes*, trans. William Horsley Gantt (New York: International Publishers, 1941), 41.
2. Karl Pearson, *The Grammar of Science* (London: Adam and Charles Black, 1900), 17.
3. John 14:6, NIV.
4. Nathan Myhrvold, quoted in "2012: What Is Your Favorite Deep, Elegant, or Beautiful Explanation?" Edge, accessed December 21, 2020, https://www.edge.org/responses/what-is-your-favorite-deep-elegant-or-beautiful-explanation.
5. René Descartes, *Discourse on the Method of Rightly Conducting the Reason and of Seeking Truth in the Sciences*, trans. John Veitch (Chicago: Open Court, 1910), 19.
6. P. W. Bridgman, *Reflections of a Physicist* (New York: Philosophical Library, 1950), 81, 83.
7. Christine V. McLelland, "Talking Points about Science," in *The Nature of Science and*

the Scientific Method, The Geological Society of America, accessed January 25, 2021, 7, https://www.geosociety.org/documents/gsa/geoteachers/NatureScience.pdf.

8. Susan Valdes, quoted in James Call, "Mental Health Days for Students: An Increasingly Accepted Reason to Stay Home from School," *USA Today*, October 19, 2019, https://www.usatoday.com/story/news/education/2019/10/19/mental-health-day-students-more-states-embracing/4036423002/.
9. See, for example, "The Effects of Technology on Mental Health," *Fisher-Titus Healthy Living* (blog), September 18, 2018, https://www.fishertitus.org/health/effects-technology-mental-health.
10. Albert Einstein, "Physics and Reality," *Journal of the Franklin Institute* 221, no. 3 (March 1936): 349–382.
11. Isaiah 55:8-9.
12. Gottfried Wilhelm Leibniz, "On the Ultimate Origination of the Universe," in *Monadology and Other Philosophical Essays*, trans. Paul Schrecker and Anne Martin Schrecker (Indianapolis: Bobbs-Merrill, 1965), 85.
13. Firas Alkhateeb, *Lost Islamic History: Reclaiming Muslim Civilisation from the Past* (London: Hurst, 2014), 78.
14. René Descartes, "Principles of Philosophy (1644)," trans. John Veitch, in *The History and Philosophy of Science: A Reader*, ed. Daniel J. McKaughan and Holly VandeWall (London: Bloomsbury, 2018), 362, emphasis added.
15. It's often spelled Occam's Razor.
16. William of Ockham, *Summa Logicae*, Part 1, chap. 12, para. 6.
17. Huston Smith, quoted in Kimberly Winston, "Huston Smith: A History of Faith," *Publishers Weekly*, January 22, 2001, https://www.publishersweekly.com/pw/by-topic/authors/interviews/article/37180-huston-smith-a-history-of-faith.html.
18. For a brief article that clears up some common misconceptions about Franklin's experiment, see Nancy Gupton, "Benjamin Franklin and the Kite Experiment," The Franklin Institute, June 12, 2017, https://www.fi.edu/benjamin-franklin/kite-key-experiment.
19. Malcolm Macleod et al., "Risk of Bias in Reports of In Vivo Research: A Focus for Improvement," PLOS Biology, October 13, 2015, https://journals.plos.org/plosbiology/article?id=10.1371/journal.pbio.1002273.
20. Monya Baker, "1,500 Scientists Lift the Lid on Reproducibility," *Nature*, May 25, 2016, https://www.nature.com/news/1-500-scientists-lift-the-lid-on-reproducibility-1.19970.
21. Margaret Mead, "The Role of the Individual in Samoan Culture," *Journal of the Royal Anthropological Institute of Great Britain and Ireland* 58 (Jul–Dec, 1928): 481–495, https://doi.org/10.2307/2843632.
22. Derek Freeman, *Margaret Mead and Samoa: The Making and Unmaking of an Anthropological Myth* (Cambridge, MA: Harvard University Press, 1983), 262.
23. Derek Freeman, "Paradigms in Collision: Margaret Mead's Mistake and What It Has Done to Anthropology," *Skeptic* 5, no. 3, 68, https://www.jstor.org/stable/j.ctt1rfsrvv.6.
24. Daniele Fanelli, "How Many Scientists Fabricate and Falsify Research? A Systematic Review and Meta-Analysis of Survey Data," PLoS ONE 4, no. 5 (May 29, 2009),

https://journals.plos.org/plosone/article?id=10.1371/journal.pone.0005738&imageURI=info:doi/10.1371/journal.pone.0005738.t001.

CHAPTER 8: HAVING FAITH IN MATHEMATICS

1. Quoted in Constance Reid, "Being Julia Robinson's Sister," in Bettye Anne Case and Anne M. Leggett, eds., *Complexities: Women in Mathematics* (Princeton, New Jersey: Princeton University Press, 2005), 18.
2. "Does mathematical proof require faith," StackExchange Philosophy, July 2, 2017, https://philosophy.stackexchange.com/questions/43388/does-mathematical-proof-require-faith.
3. Gottlob Frege, quoted in *The Frege Reader*, Michael Beaney, ed. (Malden, MA: Blackwell, 1997), 254.
4. Frege, in *The Frege Reader*, 279.
5. J. J. O'Connor and E. F. Robertson, "Friedrich Ludwig Gottlob Frege," MacTutor, November 2002, http://www-history.mcs.st-andrews.ac.uk/Biographies/Frege.html.
6. David Hilbert, "On the Infinite," in *Philosophy of Mathematics: Selected Readings*, 2nd ed., ed. Paul Benacerraf and Hilary Putnam, trans. Erna Putnam and Gerald J. Massey (Cambridge, UK: Cambridge University Press, 1983), 191.
7. My brief description here summarizes the combined results of Gödel's two incompleteness theorems. For a more in-depth discussion, see Panu Raatikainen, "Gödel's Incompleteness Theorems," *Stanford Encyclopedia of Philosophy*, April 2, 2020, https://plato.stanford.edu/entries/goedel-incompleteness/.
8. Verena Huber-Dyson, quoted in Siobhan Roberts, "Waiting for Gödel," *New Yorker*, June 29, 2016, https://www.newyorker.com/tech/annals-of-technology/waiting-for-godel. Italics in the original.
9. Morris Kline, *Mathematics for the Nonmathematician* (New York: Dover, 1985). Kline's remark appears to be a paraphrase of an observation by Joseph Wood Krutch, an early critic of modernism: "Metaphysics may be, after all, only the art of being sure of something that is not so and logic only the art of going wrong with confidence." See Joseph Wood Krutch, *The Modern Temper* (New York: Harcourt Brace, 1929), 228.
10. Bertrand Russell, "Reflections on My Eightieth Birthday," in *Portraits from Memory: And Other Essays* (New York: Simon and Schuster, 1956), 54–55, emphasis in the original.
11. John D. Barrow, *The Artful Universe: The Cosmic Source of Human Creativity* (Boston: Back Bay Books, 1995), 211.
12. Heidar A. Malki, quoted in "What is 'Fuzzy Logic'? Are There Computers That Are Inherently Fuzzy and Do Not Apply the Usual Binary Logic," *Scientific American*, October 21, 1999, https://www.scientificamerican.com/article/what-is-fuzzy-logic-are-t/.
13. Edward Nelson, "Mathematics and Faith," n.d., https://web.math.princeton.edu/~nelson/papers/faith.pdf, 1.
14. Albert Einstein, "Geometry and Experience: An Expanded Form of an Address by Albert Einstein to the Prussian Academy of Sciences in Berlin on 27 Jan 1921," Today in Science History, https://todayinsci.com/E/Einstein_Albert/Einstein-GeometryAndExperience.htm.

15. See "Eugene Wigner: Biographical," Nobel Prize, https://www.nobelprize.org/prizes/physics/1963/wigner/biographical/.
16. Eugene P. Wigner, "The Unreasonable Effectiveness of Mathematics in the Natural Sciences," in *Communications in Pure and Applied Mathematics* 13, no. 1 (Feb 1960), https://www.dartmouth.edu/~matc/MathDrama/reading/Wigner.html.
17. John Horgan, "The Death of Proof," *Scientific American*, October 1993, https://www.scientificamerican.com/article/the-death-of-proof/.
18. Keith Devlin, quoted in Roxanne Khamsi, "Mathematical Proofs Getting Harder to Verify," *New Scientist*, February 19, 2006, https://www.newscientist.com/article/dn8743-mathematical-proofs-getting-harder-to-verify/.
19. John Rath, "In Texas, A Stampede of Petaflops," Data Center Knowledge, March 28, 2013, https://www.datacenterknowledge.com/archives/2013/03/28/stampede-supercomputer-goes-live.
20. "The Longest Proof in the History of Mathematics," CNRS News, July 20, 2016, https://news.cnrs.fr/articles/the-longest-proof-in-the-history-of-mathematics.

CHAPTER 9: HAVING FAITH IN PHYSICS

1. Joe Young and Sam M. Lewis, "How Ya Gonna Keep 'em Down on the Farm" (New York: Waterson, Bernstein & Snyder, 1919).
2. Aristotle, *Metaphysics*, trans. W. D. Ross, book XII, part 8, para. 6 http://classics.mit.edu/Aristotle/metaphysics.12.xii.html.
3. Albert Einstein, "Über einen die Erzeugung und Verwandlung des Lichtes betreffenden heuristischen Gesichtspunkt" [On a heuristic point of view about the creation and conversion of light], *Annalen der Physik* 322, no. 6 (1905), 132–148.
4. Isaac Asimov and Jason A. Shulman, eds., *Isaac Asimov's Book of Science and Nature Quotations* (London: Weidenfeld and Nicolson, 1988), 327.
5. Richard P. Feynman, *The Character of Physical Law* (New York: BBC/Penguin, 1965), 129.
6. Pierre Speziali, ed., *Albert Einstein–Michele Besso Correspondence, 1903-1955* (Paris: Hermann, 1972), 453. See, for instance, https://www.spaceandmotion.com/quantum-theory-albert-einstein-quotes.htm.
7. In de Broglie's honor, we now describe a particle's momentum in terms of its *de Broglie wavelength*.
8. Tim Childers, "Physicists Link Quantum Memories across the Longest Distance eEver," Live Science, March 5, 2020, https://www.livescience.com/quantum-memory-entangled-far.html.
9. Martin Rees, "The Anthropic Universe," *New Scientist* 115, no. 1572 (August 6, 1987): 46.
10. Werner Heisenberg, "The Representation of Nature in Contemporary Physics," in Rollo May, ed., *Symbolism in Religion and Literature* (New York: George Braziller, 1960), 231.
11. Brian Greene, "What Is the Quantum Measurement Problem," World Science U, YouTube video, 0:43, April 5, 2018, https://youtu.be/nf5yiH156VM?t=6.
12. For all you photography buffs: Yes, I know that the focal length of the lens makes a difference. But I trust you get the point I'm making.
13. Albert Einstein, "Zur Elektrodynamik bewegter Körper" [On the electrodynamics of moving bodies], *Annalen der Physik* 322, no. 10 (1905): 891–921.

14. Einstein's theory of general relativity revealed that when gravity is present, a proper description of spacetime requires what's called a 4D Riemannian geometry.
15. "Penny," United States Mint, https://www.usmint.gov/coins/coin-medal-programs/circulating-coins/penny.
16. From now on, whenever I say, "the speed of light," I mean the speed of light *in a vacuum*: 299,792,458 meters per second.
17. "Original Letter from Isaac Newton to Richard Bentley," The Newton Project, October 2007, http://www.newtonproject.ox.ac.uk/view/texts/normalized/THEM00258.
18. His announcement was published on December 2, 1915, in the *Proceedings of the Royal Prussian Academy of Sciences*. See Albert Einstein, *The Collected Papers of Einstein*, vol. 6, *The Berlin Years: Writings*, 1914–1917 (Princeton, New Jersey: Princeton University Press, 1996), 244, https://einsteinpapers.press.princeton.edu/vol6-doc/272.
19. The full backstory of how Einstein came to publish his theory of general relativity is fascinating. For a good layman's account, see Sarah Pruitt, "6 Things You Might Not Know about Einstein's General Theory of Relativity," *History*, August 29, 2018, https://www.history.com/news/6-things-you-might-not-know-about-einsteins-general-theory-of-relativity.
20. Leonardo da Vinci, *The Literary Works of Leonardo da Vinci*, ed. Jean Paul Richter, vol. 1 (London: Sampson Low, Marston, Searle & Rivington, 1883), 11. Plato felt the same way as Leonardo. Reportedly, a sign over the entrance to his Academy read, "Let no one ignorant of geometry enter."
21. *The Life of Leonardo da Vinci*, season 1, episode 5, directed by Renato Castellani, aired November 21, 1971, on RAI Radiotelevisione Italiana.

CHAPTER 10: HAVING FAITH IN ASTRONOMY

1. Plato, *The Republic*, trans. B. Jowett (Project Gutenberg, June 22, 2016), Book VII.
2. Sir William Bragg, *The Universe of Light* (New York: Macmillan, 1933), 1.
3. See, for instance, "Dark Energy, Dark Matter," NASA Science, accessed December 27, 2020, https://science.nasa.gov/astrophysics/focus-areas/what-is-dark-energy.
4. *The Mystery of Dark Matter,* directed by Cécile Denjean (Arte France and Scientifilms, 2012), starting at 1:33.
5. *The Mystery of Dark Matter*, 50:36.
6. Greg Bothun, *Modern Cosmological Observations and Problems* (London: Taylor & Francis, 1998), ix, 1.
7. I'm speaking loosely here to keep the language simple. When I refer to the universe's *weight*—the total amount of its mass and energy—I mean the total amount of mass and energy *divided by the total volume of the universe*. That quotient gives us the total mass-energy density of the universe.
8. A flat universe takes an infinite amount of time to dissolve into a cold, dead nothingness. An open universe takes a finite time to do so.
9. "Flat" does not mean the universe is literally flat, like a tabletop. It means it follows the rules of ordinary Euclidean plane geometry—the kind they teach in high school.
10. Leah Crane, "Cosmological Crisis: We Don't Know If the Universe Is Round or Flat," *New Scientist*, November 4, 2019, https://www.newscientist.com/article/2222159-cosmological-crisis-we-dont-know-if-the-universe-is-round-or-flat/.

11. Wendy L. Freedman et al., "Final Results from the Hubble Space Telescope Key Project to Measure the Hubble Constant," *Astrophysical Journal* 553, no. 1 (2001): 47–72, https://ned.ipac.caltech.edu/level5/Sept01/Freedman/Freedman_contents.html.
12. Adam G. Riess et al., "Large Magellanic Cloud Cepheid Standards Provide a 1% Foundation for the Determination of the Hubble Constant and Stronger Evidence for Physics Beyond LambdaCDM," Cornell University, March 27, 2019, https://arxiv.org/abs/1903.07603.
13. Inh Jee et al., "A Measurement of the Hubble Constant from Angular Diameter Distances to Two Gravitational Lenses," *Science* 365, no. 6458 (September 13, 2019): 1134–1138, https://science.sciencemag.org/content/365/6458/1134.
14. Adam Riess, quoted in Corey S. Powell, "The Universe May Be a Billion Years Younger Than We Thought. Scientists Are Scrambling to Figure Out Why," MACH, May 18, 2019, https://www.nbcnews.com/mach/science/universe-may-be-billion-years-younger-we-thought-scientists-are-ncna1005541.
15. Jeff Glorfeld, "Georges Lemaître Comes In with a Bang," *Cosmos*, October 24, 2020, https://cosmosmagazine.com/sciences/physics/georges-lemaitre-comes-in-with-a-bang/.
16. Arthur S. Eddington, "The End of the World: From the Standpoint of Mathematical Physics," *Nature* 127 (March 21, 1931): 450.
17. BBC radio broadcast transcribed in *The Listener* (April 1949). See "Fred Hoyle: An Online Exhibition," St. John's College, University of Cambridge, http://www.joh.cam.ac.uk/library/special_collections/hoyle/exhibition/radio/.
18. Eric M. Jones, "Where is Everybody?" *Physics Today* 38, no. 8 (August 1995): 11, https://physicstoday.scitation.org/doi/abs/10.1063/1.2814654?journalCode=pto&.
19. See, for example, "What Is an Exoplanet?" NASA Science, June 4, 2020, https://spaceplace.nasa.gov/all-about-exoplanets/en/. See also "Goldilocks Zone," NASA Exoplanet Exploration, September 24, 2020, https://exoplanets.nasa.gov/resources/323/goldilocks-zone/.
20. "Are there any exoplanets like Earth?" in "The Search for Life," NASA Exoplanet Exploration, December 8, 2020, https://exoplanets.nasa.gov/search-for-life/big-questions/.
21. There is some debate about the exact definition of "natural." Thus, some argue that there are as many as ninety-eight natural elements and twenty synthetic ones.
22. Richard Dawkins, interview with Ben Stein, in *Expelled: No Intelligence Allowed*, directed by Nathan Frankowski, April 19, 2008 (Premise Media Corp., 2008), starting at 1:30:09.
23. Charles Darwin, "To J. D. Hooker 1 February [1871]," Darwin Correspondence Project, February 1, 1871, https://www.darwinproject.ac.uk/letter/DCP-LETT-7471.xml.
24. According to the American Society of Hematology, patients with sickle cell disease have "a national median life expectancy of 42–47 years." See "Rare Patients with Sickle Cell Disease Live Nearly Twice as Long as Average," American Society of Hematology, October 24, 2016, https://www.hematology.org/newsroom/press-releases/2016/rare-patients-with-sickle-cell-disease-live-nearly-twice-as-long-as-average.
25. Elena A. Ponomarenko et al., "The Size of the Human Proteome: The Width and Depth," *International Journal of Analytical Chemistry*, May 19, 2016, https://www.ncbi.nlm.nih.gov/pmc/articles/PMC4889822/.

26. Sir Fred Hoyle and Chandra Wickramasinghe, *Evolution from Space* (New York: Touchstone, 1984), 3.
27. Hoyle and Wickramasinghe, *Evolution from Space*, 30.
28. Ilya Prigogine, Gregoire Nicolis, and Agnes Babloyantz, "Thermodynamics of Evolution," *Physics Today* 25, no. 12 (December 1972): 38.
29. Prigogine, et al., "Thermodynamics of Evolution."
30. This section is adapted from an op-ed piece I wrote for Fox News in 2018. See Michael Guillen, "It's Quite Possible to Believe in 'Little Green Men' but It's Just Gotten Harder," Fox News, July 21, 2018, https://www.foxnews.com/opinion/its-quite-possible-to-believe-in-little-green-men-but-its-just-gotten-harder.
31. Anders Sandberg, Eric Drexler, and Toby Ord, "Dissolving the Fermi Paradox," Future of Humanity Institute, Oxford University, June 8, 2018, https://arxiv.org/pdf/1806.02404.pdf.
32. Sandberg et al., "Dissolving the Fermi Paradox," 16.
33. Sandberg et al., "Dissolving the Fermi Paradox," 16.

CHAPTER 11: HAVING FAITH IN GOD

1. Andrew McChesney, "Propelled by TMI, Adventist Church Tops 20 Million Members," *Adventist Review*, March 1, 2017, https://www.adventistreview.org/church-news/story-propelled-by-tmi,-adventist-church-tops-20-million-members.
2. William Blake, "The Everlasting Gospel," in *The Oxford Book of English Mystical Verse*, ed. D. H. S. Nicholson and A. H. E. Lee (Oxford: Clarendon Press, 1924), 94. For the entire text of the poem online, see https://www.bartleby.com/236/58.html.
3. "Number of English Translations of the Bible," American Bible Society News, December 2, 2009, http://news.americanbible.org/article/number-of-english-translations-of-the-bible.
4. For a list of the apocryphal books and links to the texts, see "New Testament Apocrypha: The Missing Books of the Bible," Interfaith Online, https://www.interfaith.org/christianity/apocrypha/. For an excellent critique of the Apocrypha, see Don Stewart, "Why Were the Books of the Old Testament Apocrypha Rejected as Holy Scriptures by the Protestants?" Blue Letter Bible, https://www.blueletterbible.org/faq/don_stewart/don_stewart_395.cfm.
5. See, for example, Graham P. Collins, "The Many Interpretations of Quantum Mechanics," *Scientific American*, November 19, 2007, https://www.scientificamerican.com/article/the-many-interpretations-of-quantum-mechanics/.
6. If you want a refresher in why quantum physics is beyond proof, review chapter 7, "Having Faith in the Scientific Method," and chapter 8, "Having Faith in Mathematics."
7. Albert Einstein, quoted in Scott Bembenek, "Einstein and the Quantum," *Scientific American* (blog), March 27, 2018, https://blogs.scientificamerican.com/observations/einstein-and-the-quantum/.
8. Michael Guillen, *Amazing Truths: How Science and the Bible Agree* (Zondervan, 2016).
9. Lawrence Mykytiuk, "53 People in the Bible Confirmed Archaeologically," Bible History Daily, October 13, 2020, https://www.biblicalarchaeology.org/daily/people-cultures-in-the-bible/people-in-the-bible/50-people-in-the-bible-confirmed-archaeologically/. See also Lawrence Mykytiuk, "Archaeology Confirms 50 Real People in the Bible," *Biblical Archaeological Review* 40, no. 2 (March/April 2014);

Lawrence Mykytiuk, "Archaeology Confirms 3 More Bible People," *Biblical Archaeological Review* 43, no. 3 (May/June 2017).
10. See, for example, J. Barton Payne, *Encyclopedia of Biblical Prophecy: The Complete Guide to Scriptural Predictions and Their Fulfilment* (Baker, 1980).
11. See, for example, "Did Jesus Fulfill the Messianic Prophecies in the Old Testament?" Jewish Voice, https://www.jewishvoice.org/learn/jesus-did-not-fulfill-messianic-prophecies-found-old-testament.
12. See, for example, Jonathan Petersen, "When Was Each Book of the Bible Written?" *Bible Gateway Blog*, February 1, 2016, https://www.biblegateway.com/blog/2016/02/when-was-each-book-of-the-bible-written/.
13. Isaiah 7:14.
14. Micah 5:2.
15. Zechariah 9:9.
16. Zechariah 12:10.
17. Isaiah 53:3-9.
18. Daniel 9:26.
19. Daniel 7:13-14.
20. Isaiah 2:4.
21. Luke 1:31, NIV.
22. Luke 1:34, NIV.
23. Isaiah 7:14.
24. See Isaiah 7:14, "Verse-by-Verse Bible Commentary," StudyLight.org, https://www.studylight.org/commentary/isaiah/7-14.html.
25. Isaiah 7:14, "Verse-by-Verse."
26. See, for example, Leah Lefler, "Parthenogenesis: Virgin Births in Nature," Owlcation, August 19, 2012, https://owlcation.com/stem/Parthenogenesis-Virgin-Births-in-Nature.
27. Warren Booth, quoted in Melissa Hogenboom, "Spectacular Real Virgin Births," BBC Earth, December 22, 2014, http://www.bbc.com/earth/story/20141219-spectacular-real-virgin-births.
28. Gabriel Jose de Carli and Tiago Campos Pereira, "On Human Parthenogenesis," *Medical Hypotheses* 106 (September 2017): 57–60, https://www.sciencedirect.com/science/article/pii/S0306987717302694.
29. Sylvia Pagán Westphal, "'Virgin Birth' Mammal Rewrites Rules of Biology," *New Scientist*, April 21, 2004, https://www.newscientist.com/article/dn4909-virgin-birth-mammal-rewrites-rules-of-biology/.
30. See, for example, Lawrence Mykytiuk, "Did Jesus Exist? Searching for Evidence Beyond the Bible," Bible History Daily, October 2, 2020, https://www.biblicalarchaeology.org/daily/people-cultures-in-the-bible/jesus-historical-jesus/did-jesus-exist/; Don Stewart, "What Do Early Non-Christian Writings Say about Jesus?" Blue Letter Bible, https://www.blueletterbible.org/faq/don_stewart/don_stewart_185.cfm; and Michael Gleghorn, "Ancient Evidence for Jesus from Non-Christian Sources," Probe, August 30, 2014, https://probe.org/ancient-evidence-for-jesus-from-non-christian-sources-2/.
31. Bart Ehrman, "Non-Christian Sources for Jesus: An Interview with History.com," *The Bart Ehrman Blog*, February 24, 2019, https://ehrmanblog.org/non-christian-sources-for-jesus-an-interview-with-history-com/.
32. Bart Ehrman, "With Respect to Others Who Did Not Like My *Newsweek* Article,"

The Bart Ehrman Blog, January 2, 2021, https://ehrmanblog.org/with-respect-to-others-who-did-not-like-my-newsweek-article/.

33. Flavius Josephus, *The Antiquities of the Jews*, trans. William Whiston, vol. 18, chap. 3, para. 3, Project Gutenberg, 2009, https://www.gutenberg.org/files/2848/2848-h/2848-h.htm.
34. Josephus, *Antiquities*, vol. 20, chap. 9, para. 1.
35. Peter Shäfer, *Jesus in the Talmud* (Princeton, New Jersey: Princeton University Press, 2007), 10. The Talmud is an ancient, massive, centuries-long compilation of Jewish law and commentary.
36. Lawrence Mykytiuk, quoted in Christopher Klein, "The Bible Says Jesus Was Real. What Other Proof Exists?" History.com, February 26, 2019, https://www.history.com/news/was-jesus-real-historical-evidence.
37. From research conducted in 2015 and 2016 on behalf of the Church of England, Evangelical Alliance, and HOPE. See Lucy Olofinjana and Catherine Butcher, *Talking Jesus: Dig Deeper—What People in England Think of Jesus, Christians and Evangelism (2018)*, https://talkingjesus.org/wp-content/uploads/2018/04/Talking-Jesus-dig-deeper.pdf.
38. I say this because Christianity is the largest, most widespread belief system in the world. See, for example, Conrad Hackett and David McClendon, "Christians Remain World's Largest Religious Group, but They Are Declining in Europe," Fact Tank, Pew Research Center, April 5, 2017, https://www.pewresearch.org/fact-tank/2017/04/05/christians-remain-worlds-largest-religious-group-but-they-are-declining-in-europe/.
39. See, for example, Pallavi Thakur, "How Did Lord Buddha Die?" Speaking Tree, May 26, 2017, https://www.speakingtree.in/allslides/how-did-lord-buddha-die/264131 and "Buddha Biography" Biography, updated July 13, 2020, https://www.biography.com/religious-figure/buddha.
40. Jesus' death is reported not only in the New Testament but also by other reliable, independent sources. For example, in *The Annals*, Tacitus, a non-Christian Roman historian reports concerning the fire in Rome of 64 AD: "Nero fastened the guilt and inflicted the most exquisite tortures on a class hated for their abominations, called Christians by the populace. Christus, from whom the name had its origin, suffered the extreme penalty during the reign of Tiberius at the hands of one of our procurators, Pontius Pilatus" (Tacitus, *The Annals*, vol. 15, trans. Alfred John Church and William Jackson Brodribb, http://classics.mit.edu/Tacitus/annals.11.xv.html). Also, Sextus Julius Africanus, a third-century Christian historian, recounts the circumstances of Jesus' crucifixion: "On the whole world there pressed a most fearful darkness; and the rocks were rent by an earthquake, and many places in Judea and other districts were thrown down. This darkness Thallus, in the third book of his *History*, calls, as appears to me without reason, an eclipse of the sun." See Fragment 18, para. 1 in Sextus Julius Africanus, "The Extant Fragments of the Five Books of the Chronography of Julius Africanus," JasonColavito.com, http://www.jasoncolavito.com/julius-africanus-chronography.html.
41. John 15:13, CEV.
42. Genesis 3:16.
43. Genesis 3:17-19.

44. Genesis 6:9.
45. Genesis 7:11-12.
46. See, for example, Tzvi Freeman, "Tikkun Olam," Chabad.org, https://www.chabad.org/library/article_cdo/aid/3591946/jewish/Tikkun-Olam.htm.
47. Polls consistently show that Christians—along with Jews and other religious adherents—are more philanthropic than Atheists. "People who are religiously affiliated are more likely to make a charitable donation of any kind," reports Giving USA, "whether to a religious congregation or to another type of charitable organization." See "Giving USA Special Report on Giving to Religion," Giving USA, October 24, 2017, https://givingusa.org/just-released-giving-usa-special-report-on-giving-to-religion/.
48. John 14:6, emphasis added.
49. See John 14:6; Acts 9:2; 18:25; 19:9, 23; 22:4; 24:14, 22.
50. 1 Corinthians 15:14-15.
51. 1 Corinthians 15:5-8, emphasis added.
52. Acts 8:1, 3.
53. See, for example, James Tabor, "The Quest for the Historical Paul," Bible History Daily, Biblical Archaeology Society, November 1, 2020, https://www.biblicalarchaeology.org/daily/people-cultures-in-the-bible/people-in-the-bible/the-quest-for-the-historical-paul/#note07r.
54. Philippians 3:6.
55. Galatians 1:13.
56. The same problem exists for other quasars as well: How could they have become so massive in such a short time? For a discussion of the problem, see, for example, Leah Crane, "Most Distant Quasar Ever Seen Is Way Too Big for Our Universe," *New Scientist*, December 6, 2017.
57. E. P. Sanders, *The Historical Figure of Jesus* (New York: Penguin Books, 1995), 10–11, 280.
58. Gary R. Habermas and Michael R. Licona, *The Case for the Resurrection of Jesus* (Grand Rapids, MI: Kregel, 2004), 44.
59. "Gary Habermas: The Resurrection Evidence That Changed Current Scholarship," BiolaUniversity, YouTube video, 1:14:30, September 14, 2013, https://youtu.be/5znVUFHqO4Q?t=1714, starting at 28:34.
60. Romans 1:20.
61. John 3:16, my definition and emphasis added.
62. *Merriam-Webster*, s.v. "trust (*n.*)," accessed February 20, 2021, https://www.merriam-webster.com/dictionary/trust.
63. 1 Corinthians 2:14, NIV.
64. Matthew 11:28-29.
65. Matthew 16:25, ESV.
66. John 11:25, GWT.
67. Galatians 1:11-12, 15-16.
68. 2 Corinthians 5:17.
69. See, for example, D. D. Emmons, "From Persecutor to Christian: The Conversion of St. Paul," Our Sunday Visitor, January 25, 2015, https://osvnews.com/2015/01/25/from-persecutor-to-christian-the-conversion-of-st-paul/.

70. 2 Corinthians 11:24-30, 12:7-10, NIV.
71. See Acts 9:1-30.
72. His name was the Reverend Doctor Juan Hernandez. He was a faithful and precious friend to me and my family—and a true man of God.
73. 1 Corinthians 2:13.
74. See proposition 5.6 in Ludwig Wittgenstein, *Tractatus Logico-Philosophicus*, Ogden translation (London: Kegan Paul, 1922), 533, http://writing.upenn.edu/library /Wittgenstein-Tractatus.pdf.
75. Abraham Joshua Heschel, *Man Is Not Alone: A Philosophy of Religion* (New York: Farrar, Straus and Giroux, 1976), 25.
76. 1 John 4:18, ESV.
77. John 11:25, GWT.
78. 1 John 4:16-17.
79. See, for example, Romans 1:28-32.
80. See, for example, Romans 1:19-20.
81. Michael Lipka, "10 Facts about Atheists," Fact Tank, Pew Research Center, December 6, 2019, https://www.pewresearch.org/fact-tank/2019/12/06/10-facts-about-atheists/.
82. Staks Rosch, "Atheists Can Be Spiritual Too," *HuffPost* (blog), updated December 6, 2017, https://www.huffpost.com/entry/atheists-can-be-spiritual_b_1316619.
83. Sam Harris, "The Problem with Atheism," *Sam Harris* (blog), October 2, 2007, https://samharris.org/the-problem-with-atheism/.
84. Sam Harris, *Waking Up: A Guide to Spirituality without Religion* (Simon & Schuster, 2014).
85. Publisher's summary of Sam Harris, *Waking Up: A Guide to Spirituality without Religion* (Simon & Schuster, 2014), on amazon.com, https://www.amazon.com /Waking-Up-Spirituality-Without-Religion/dp/1451636016/ref=tmm_hrd_swatch_0? _encoding=UTF8&qid=&sr=.
86. Genesis 2:9.
87. Genesis 2:17.
88. See, for example, C. R. Nave, "Asymptotic Freedom," *HyperPhysics*, Georgia State University, 2016, http://hyperphysics.phy-astr.gsu.edu/hbase/Particles/qbag.html.
89. 1 Corinthians 2:14, BSB.
90. 1 Corinthians 2:11-13.
91. Matthew 22:37-38.
92. Genesis 2:17, NIV.
93. John 10:30.
94. John 10:31.
95. John 10:32-33.
96. John 14:25-26, NIV.
97. Luke 17:20-21, GNT.
98. John 14:26.
99. John 8:32, NIV.

CHAPTER 12: HAVING FAITH IN YOUR WORLDVIEW

1. "Gender Identity: Can a 5'9, White Guy Be a 6'5, Chinese Woman?" Family Policy Institute of Washington, YouTube video, 4:13, https://www.youtube.com/watch?v =xfO1veFs6Ho.

2. Hans Christian Andersen, "The Emperor's New Clothes," bartleby.com, https://www.bartleby.com/17/3/3.html.
3. "Well-Known Expressions: The Emperor Has No Clothes," Book Browse, https://www.bookbrowse.com/expressions/detail/index.cfm/expression_number/605/the-emperor-has-no-clothes.
4. Gregory Hoyt, "Convicted Sex Offender: I Identify as an 8-year-old. Child Porn Is My Constitutional Right," Law Enforcement Today, January 12, 2020, https://www.lawenforcementtoday.com/convicted-sex-offender-i-identify-as-an-8-year-old-child-porn-is-my-constitutional-right/.
5. See, for example, Rick Maese, "Stripped of Women's Records, Transgender Powerlifter Asks, 'Where Do We Draw the Line?'" *Washington Post*, May 19, 2019, https://www.washingtonpost.com/sports/2019/05/16/stripped-womens-records-transgender-powerlifter-asks-where-do-we-draw-line/. See also Christie Aschwanden, "Trans Athletes Are Posting Victories and Shaking Up Sports," Wired, October 29, 2019, https://www.wired.com/story/the-glorious-victories-of-trans-athletes-are-shaking-up-sports/.
6. "Most Oppose Transgender Athletes on Opposite Sex Teams," Rasmussen Reports, June 4, 2019, https://www.rasmussenreports.com/public_content/lifestyle/social_issues/most_oppose_transgender_athletes_on_opposite_sex_teams.
7. Jacob Gershman, "States Weigh Measures to Stop Transgender Athletes from Competing in Women's Sports," *Wall Street Journal*, January 7, 2020, https://www.wsj.com/articles/states-weigh-measures-to-stop-transgender-athletes-from-competing-in-womens-sports-11578393001. See also Samantha Pell, "Girls Say Connecticut's Transgender Athlete Policy Violates Title IX, File Federal Complaint," *Washington Post*, June 19, 2019, https://www.washingtonpost.com/sports/2019/06/19/girls-say-connecticuts-transgender-athlete-policy-violates-title-ix-file-federal-complaint/.
8. "The Global Religious Landscape," Pew Research Center, December 18, 2012, https://www.pewforum.org/2012/12/18/global-religious-landscape-exec/.
9. George Gallup and Timothy K. Jones, *The Next American Spirituality: Finding God in the Twenty-first Century*, (Colorado Springs: Cook Communications, 2000), 58.
10. "Icebergs in the North Atlantic Ocean," National Geographic Resource Library, updated April 9, 2012, https://www.nationalgeographic.org/media/iceberg-frequency/.
11. Martin Luther King, Jr., "I Have a Dream," speech at the Lincoln Memorial, Washington, DC, August 28, 1963. Transcript on American Rhetoric: Top 100 Speeches, https://www.americanrhetoric.com/speeches/mlkihaveadream.htm.
12. See, for example, "Flow" *Psychology Today*, https://www.psychologytoday.com/us/basics/flow.
13. Owen Amos, "Did Martin Luther King Predict His Own Death?" BBC News, March 26, 2018, https://www.bbc.com/news/world-us-canada-43545620.
14. Psalm 46:10.
15. 1 Corinthians 2:9.
16. 1 Corinthians 2:10-16, emphasis added.

CHAPTER 13: PUTTING YOUR FAITH TO THE TEST

1. This baseline number, though somewhat arbitrary, reflects the fact that even if your reality lines up perfectly with objective reality, there is still a nonzero chance you will

make a decision that harms or kills you. Using 125 as the baseline means that even in this optimal situation, you still have about a 25 percent chance of making an unwise decision—midway between an ideal 0 percent and a random 50 percent. In other words, even with an ideal worldview, you will behave not randomly or perfectly but somewhere in between.

2. Since then, other such mirrors have been placed on the moon's surface.
3. Elizabeth Howell, "Why is the Apollo Reflector Experiment Still Operating, 50 Years Later?" Space.com, July 11, 2019, https://www.space.com/apollo-retroreflector-experiment-still-going-50-years-later.html.
4. "Gaia Fact Sheet," European Space Agency, updated November 3, 2020, https://sci.esa.int/web/gaia/-/47354-fact-sheet.

INDEX

INDEX

ACKNOWLEDGMENTS

First and foremost, I thank truth and the Author of truth for busting me out of the small, imprisoning worldview of my youth. They are why I wrote—why I *could* write—this book. *Soli Deo gloria.*

I thank Laurel, my wife of thirty years—my wise, brave, and plucky comrade in arms. She read every word of the manuscript more than once and gave me excellent advice. I did not always accept it graciously, for which I am truly sorry. But I did always end up heeding it, thereby improving the book immeasurably.

I thank Dr. Rice Broocks, author of the international bestseller *God's Not Dead*, which inspired a series of successful movies. I have had the enormous pleasure of sharing the stage with him at college events worldwide. After listening to me speak about faith, he urged me to memorialize my thoughts in a book. You are now holding it your hands.

I thank Wes Yoder, my literary agent. Our journeys joined up nearly fifteen years ago at a dinner in Washington, DC, for members of the media. I could not wish for a more dedicated, respected, talented representative.

I thank Jan Long Harris, my publisher. For many years, she and I have been waiting patiently for an opportunity to work together—and finally it happened! Working alongside her to prepare this important book for publication has been a complete, long-overdue joy.

I thank Dave Lindstedt, truly one of the most gifted editors I have ever known. He and his diligent, keen-eyed colleagues—copy editors Sam Michel and Lauren Lindemulder—are without question one of the brightest constellations in the editorial firmament.

And, finally, I thank you, dear reader and fellow traveler. Life is a profound mystery, and we are all puzzling through it together. I pray that this book will help shed some light along your path. I look forward to hearing from you.

God bless you, one and all.

ABOUT THE AUTHOR

Dr. Michael Guillen, a three-time Emmy Award winner, bestselling author, and former Harvard University physics instructor, is known and loved by millions as the ABC News science editor, a post he filled for fourteen years (1988–2002). In that capacity, he appeared regularly on *Good Morning America*, *20/20*, *Nightline*, and *World News Tonight*. He is also the host of *Where Did It Come From?* a popular thirteen-part series for the History Channel.

Dr. Guillen produced *Little Red Wagon*, the award-winning motion picture written by Patrick Sheane Duncan (*Mr. Holland's Opus*, *Courage Under Fire*) and directed by David Anspaugh (*Rudy*, *Hoosiers*).

His articles have appeared in distinguished publications such as *Science News*, *Psychology Today*, the *New York Times*, and the *Washington Post*. He is a columnist for Fox News, *US News & World Report*, and the *Christian Post*.

He is the bestselling author of *Bridges to Infinity: The Human Side of Mathematics*; *Five Equations That Changed the World: The Power and Poetry of Mathematics*; *Can a Smart Person Believe in God?*; *Amazing Truths: How Science and the Bible Agree*; *The Null Prophecy* (thriller novel), and *The End of Life as We Know It: Ominous News from the Frontiers of Science*.

Dr. Guillen earned a bachelor's in physics and mathematics from UCLA; a master's in experimental physics from Cornell, and a three-dimensional doctorate in physics, mathematics, and astronomy from Cornell. He has also been awarded honorary doctorates from the University of Maryland and Pepperdine University. In 2000, he was elected to the renowned, century-old Explorers Club. He is a member of SAG-AFTRA and the Writers Guild of America.

Dr. Guillen is the host of *Science + God with Dr. G*, a popular podcast on Spotify, iTunes, Google Play, the AccessMore app, and Accessmore.com. He is also the host of *Going Viral* and other vodcast series on YouTube. See youtube.com/c/michaelguillenphd.

He is a sought-after speaker on college campuses, including God's Not Dead events with Dr. Rice Broocks.

Dr. Guillen is chairman and president of Spectacular Science Productions and the Spectacular Science Foundation.

He currently lives on a ranch in the Dallas-Fort Worth area with his wife, Laurel, and a lovable assortment of barnyard animals. For more, see michaelguillen.com.